SAE International's Dictionary of Commercial Vehicles

SAE International's Dictionary of Commercial Vehicles

JON M. QUIGLEY
WES CHOMINSKY

SAE INTERNATIONAL®

Warrendale, Pennsylvania, USA

400 Commonwealth Drive
Warrendale, PA 15096-0001 USA
E-mail: CustomerService@sae.org
Phone: 877-606-7323 (inside USA and Canada)
724-776-4970 (outside USA)
FAX: 724-776-0790

Library of Congress Catalog Number 2024936820
http://dx.doi.org/10.4271/9781468607895

ISBN-Print 978-1-4686-0788-8
ISBN-PDF 978-1-4686-0789-5
ISBN-epub 978-1-4686-0790-1

To purchase bulk quantities, please contact: SAE Customer Service

E-mail: CustomerService@sae.org
Phone: 877-606-7323 (inside USA and Canada)
724-776-4970 (outside USA)
Fax: 724-776-0790

Visit the SAE International Bookstore at books.sae.org

Publisher
Sherry Dickinson Nigam

Product Manager
Amanda Zeidan

Production and Manufacturing Associate
Michelle Silberman

Dedication

I do not think I can give so many people the thank you they deserve. I have occasionally done this in my writing, but it is impossible to overstate these people's influence. This starts with my childhood, living in a trailer and being a latchkey kid. My mom worked, and my dad always deployed somewhere in the world (he was in the Special Forces Group). Besides my family, to which I also owe an outstanding debt, the Olive and Mendoza families significantly impacted my life. I practically lived there, and they treated me like their children, the good and the not-so-good when I was not so good. I owe their children, also for their friendship growing up and for taking care of me when I was not doing so well on my own. Playing sports and listening to music with my family taught me a lot. Both are part of why I have made it where I am. Sports require discipline, which has largely influenced my life. Early on, I wanted to be a professional football player; I would play pickup games when doing something exceptional; I would refer to myself as Jack Youngblood and Lawrence McCutcheon. I played bass and wanted to be John Entwistle and Steve Harris. I would play music with Dave and Harold. In both cases, I still played when it became clear I did not have that talent.

I have worked in tobacco fields, barns, and fast-food restaurants. One day, I was with Harold when he was talking to his friend Danny. Danny was completing his engineering degree and thought I could at least make bass effects pedals if I were an engineer. This was the beginning of finding what I needed to do for a living. So, what does all this mean? I do not know, but one never knows the things or events that will impact one's life. If where I am is anything like success, I owe much to these folks and many more, and I have genuinely been fortunate. No matter where you start, you never know what things you do or experiences you have and how these significantly impact your life. This book is dedicated to those who have gotten me where I am and continue to help me find my way and to those who are still trying to see their way.

Contents

Preface

Welcome to the world of commercial vehicles, where the language of transport and logistics comes alive! This Dictionary of Commercial Vehicles is designed to be your comprehensive guide, providing a wealth of information and insights into the terminology and concepts defining commercial transportation's dynamic realm.

As industries evolve and technology continues to shape how we move goods and people, understanding the language of commercial vehicles becomes increasingly vital. Whether you are a seasoned professional in the transportation industry, a fleet manager, an enthusiast, or someone venturing into the world of logistics, this dictionary is tailored to meet your needs.

Inside these pages, you will find an extensive compilation of terms related to commercial vehicles, encompassing everything from vehicle types and components to industry regulations and emerging trends. Our goal is to empower you with knowledge, making it easier to navigate the complex landscape of commercial transportation and communicate effectively within this specialized field.

We have endeavored to make this dictionary user-friendly, providing clear and concise definitions alongside relevant examples and illustrations. The entries cover various topics, including vehicle classifications, engine technologies, safety standards, and logistics management. Additionally, we have incorporated cross-references to help you connect related concepts and deepen your understanding of the interconnected world of commercial vehicles.

Whether you want to enhance your proficiency, expand your vocabulary, or satisfy your curiosity about the machinery that keeps economies moving, this Dictionary of Commercial Vehicles is your go-to resource. As the transportation industry continues to evolve, we invite you to embark on a journey of discovery within these pages, where knowledge meets the road.

Introduction

In the pulsating heart of global commerce, the language of transportation echoes with precision, efficiency, and innovation. As we navigate the intricate network of roads, highways, and logistical challenges, a comprehensive understanding of commercial vehicles becomes increasingly paramount. Welcome to the Dictionary of Commercial Vehicles—an indispensable guide that unravels the intricate tapestry of terminology woven into commercial transportation.

This dictionary is a testament to the industry's dynamic nature, capturing the essence of innovation, sustainability, and safety that propels commercial vehicles forward. Whether you are a seasoned professional, a fleet manager optimizing routes, a farmer planning and planting, or an enthusiast fascinated by the machinery that moves our economies, this dictionary is crafted to be your beacon through the vast and ever-evolving landscape of commercial vehicles.

Within these pages, you will find an extensive lexicon meticulously curated to cover various topics ranging from vehicle classifications, advanced technologies, and safety protocols to regulatory frameworks and emerging trends. Our aim is to empower you with knowledge, enable effective communication, and foster a deeper understanding of the multifaceted world of commercial vehicles.

A lexicon refers to a collection or a list of words and their meanings within a specific language or domain. It serves as a comprehensive repository of the vocabulary used by speakers of that language or professionals within a particular field of study. Lexicons can vary in size and scope, ranging from simple dictionaries that cover everyday words to extensive specialized lexicons for technical jargon in specific industries.

In linguistics, a lexicon is considered one of the critical components of a language's structure, along with grammar and phonology. It is an essential resource for understanding and using a language effectively, as it provides definitions, pronunciations, and sometimes additional information about words, such as part of speech, usage examples, and etymology.

As we embark on this journey together, we invite you to delve into the rich tapestry of terms and concepts that define the language of transportation. Whether you are deciphering the intricacies of engine technologies, exploring the nuances of logistics management, or staying abreast of the latest industry standards, this dictionary is your compass in the vast and ever-changing landscape of commercial vehicles.

List of Acronyms

4WD	Four-wheel drive
6WD	Six-wheel drive
ABS	Anti-lock braking system
AC	Alternating current
ACC	Adaptive cruise control
ADAS	Advanced driver assistance system
ADB	Air disk brakes
ADEMS	Advanced diesel engine management system
AEB	Automatic emergency braking
Ah	Ampere hour
AI	Artificial intelligence
AMT	Automated manual transmission
AOH	Air-over-hydraulic
ASC	Automatic stability control
ATA	American Trucking Association
ATC	Automatic traction control
AWD	All-wheel drive
BBC	Bumper back of cab
BDC	Bottom dead center
BHP	Brake horsepower
BSD	Blind spot detection
BSM	Blind spot monitoring
CAC	Charge air cooler
CBE	Cab beside engine
CCA	Cold cranking amps
CEEMAT	Converter enhanced, electronically managed, automatic transmission
CFE	Cab forward engine
CIT	Cash in transit
CMVSR	Canadian Motor Vehicle Safety Standard
CNG	Compressed natural gas
CO	Carbon monoxide

CO_2	Carbon dioxide
COE	Cab over engine
CPM	Cost per mile
CTIS	Central tire inflation system
CVAR	Commercial vehicle autonomous routing
CVT	Continuous variable transmissions
DC	Direct current
DCT	Dual-clutch transmission
DEF	Diesel exhaust fluid
DF1	Diesel fuel #1
DF2	Diesel fuel #2
DOC	Diesel oxidation catalyst
DOF	Degrees of freedom
DPF	Diesel particulate filter
DSG	Direct-shift gearbox
DTC	Diagnostic trouble codes
ECM	Electronic control module
ECM	Engine control module
ECR	Engine compression release
ECU	Electronic control unit
ECU	Engine control unit
EDR	Event data recorder
EGR	Exhaust gas recirculation
ELD	Electronic logging device
EPA	Environmental Protection Agency
ESC	Electronic stability control
ESP	Electronic stability program
EV	Electric vehicle
FCW	Forward collision warning
FMCSA	Federal Motor Carrier Safety Administration
FMS	Fleet management system
FWD	Front-wheel drive
GAWR	Gross axle weight rating
GCWR	Gross combination weight rating
GLONASS	Globalnaya Navigazionnaya Sputnikovaya Sistema
GNSS	Global Navigation Satellite System
GPS	Global Positioning System
GVW	Gross vehicle weight
GVWR	Gross vehicle weight rating
HC	Hydrocarbon
HD OBD	Heavy-duty onboard diagnostics
HEUI	Hydraulically actuated electronic unit injection
HF	High frequency

HOS	Hours of service
HP	Horsepower
ICE	Internal combustion engine
ISO	International Organization for Standardization
KS	Knock sensor
LDW	Lane departure warning
LEDs	Light-emitting diodes
LF	Low frequency
LKA	Lane keeping assist
LNG	Liquefied natural gas
LPG	Liquefied petroleum gas
LSD	Limited-slip differential
N_2	Nitrogen
N_2O	Nitrous oxide
NH_3	Ammonia
NO_x	Nitrogen oxide
OEM	Original equipment manufactures
OTA	Over-the-air
PIN	Postal index number
PM	Particulate matter
PTO	Power take-off
PVB	Polyvinyl butyral
RCTA	Rear cross traffic alert
RFID	Radio frequency identification
RGN	Removable gooseneck trailer
RODS	Record of duty status
ROPS	Rollover protection structures
RPM	Revolutions per minute
RV	Recreational vehicle
SCR	Selective catalytic reduction
SOC	State of charge
SUV	Sport utility vehicle
TCS	Traction control systems
TDC	Top dead center
TMC	Technology Maintenance Council
TMS	Transportation management system
TPMS	Tire pressure monitoring system
UHF	Ultra-high frequency
V2V	Vehicle-to-vehicle
VIN	Vehicle's identification number
VOC	Volatile organic compounds
VSC	Vehicle stability control
WMS	Warehouse management system

A

4-Stroke Cycle Diesel Engine

A 4-stroke cycle diesel engine is also known as a compression ignition engine, and it is an internal combustion engine (ICE) commonly used in vehicles, industrial equipment, and power generation. It operates based on four phases or strokes within a complete engine cycle: intake, compression, power, and exhaust. This engine type is widely used in various applications due to its efficiency and robustness.

Here is a breakdown of each stroke within the 4-stroke cycle diesel engine:

1. Intake Stroke: During the intake stroke, the engine's intake valve opens, allowing air to enter the combustion chamber. Unlike gasoline engines, diesel engines do not mix fuel with air during this stroke.
2. Compression Stroke: In this stroke, the intake and exhaust valves are closed. The piston moves upward, compressing the air within the combustion chamber. The compression ratio in diesel engines is relatively high, resulting in significant air compression.
3. Power Stroke (Combustion): A fuel injector sprays a fine mist of diesel fuel directly into the highly compressed air at the top of the compression stroke. The heat generated by the compressed air causes the diesel fuel to ignite spontaneously. This fuel combustion creates a high-pressure gas that rapidly expands, driving the piston down. This downward motion of the piston is what generates power.
4. Exhaust Stroke: After the power stroke, the exhaust valve opens, and the piston moves back up the cylinder. This action pushes out the exhaust gases generated during combustion, preparing the cylinder for the next intake stroke.

The 4-stroke cycle diesel engine is renowned for its fuel efficiency and durability, making it suitable for heavy-duty applications where torque and efficiency are crucial. It is commonly used in commercial trucks, buses, construction equipment, ships, and generators. Compression ignition refers to the ignition process where fuel ignites due to the high temperature resulting from the air's compression, unlike gasoline engines, where a spark plug ignites the fuel.

ABS (Anti-Lock Brake System)

ABS is an advanced automotive safety feature designed to enhance the braking performance of commercial vehicles and other automobiles (Figure A.1). It prevents the wheels of a vehicle from locking up during heavy braking or in slippery road conditions, allowing the driver to maintain steering control while reducing the risk of skidding and loss of traction [1].

FIGURE A.1 An example of an ABS architecture.

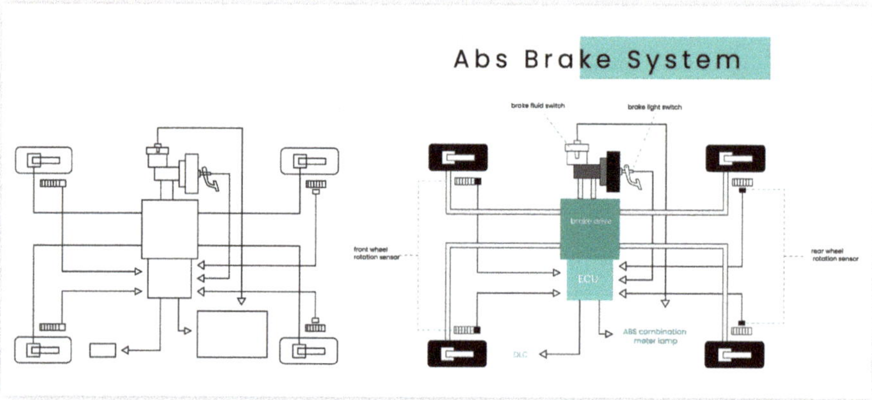

yazid_ilyas/Shutterstock.com.

For more information, see **J1939** [2].

ABS Truck Hydraulic Control Unit

This unit controls brake fluid pressure to the wheels during an anti-lock stop. It maintains, decreases, or increases the braking force as needed at each affected wheel. It is called a brake modulator on air ABS.

For more information on this, see **J2627_200908** [3].

ABS Truck Malfunction System

Medium- and heavy-duty trucks and buses are required to have ABS under Federal Motor Vehicle Safety Standards (FMVSS) 121 and must use ABS malfunction indicator lamps (telltale). These lamps must be yellow and illuminate when the ABS management controller senses a malfunction that affects the generation or transmission of response or control signals from the wheel speed sensors. ABS malfunction indicator lamps do not illuminate for all malfunction codes, only specific safety-related codes. These lamps are required to illuminate for about a 5-sec bulb check when the ignition or power switch is engaged. The warning lamps for tractor–trailers and dollies are not required to illuminate for a bulb check unless the vehicle is stopped.

For more information on this, see **J2627_200908** [3] and **FMVSS 101** controls and displays.

ACC (Adaptive Cruise Control)

ACC refers to an advanced driver assistance system (ADAS) designed to enhance the driving experience and safety of commercial vehicles. ACC combines traditional cruise control functionality with sensors and radar technology to adjust the vehicle's speed to maintain a safe following distance from the vehicle ahead.

Acceleration

Acceleration is the rate at which the velocity of an object changes over time, typically measured in meters per second squared (m/s^2) (Figure A.2).

FIGURE A.2 Acceleration in various locations on a heavy vehicle [4].

number	position	direction
1, 2	Front wheel center (left/right)	X,Y,Z/X,Z
3, 4	Middle wheel center (left/right)	X,Y,Z/X,Z
5, 6	Rear wheel center (left/right)	X,Y,Z/X,Z
7, 8	Front Engine support (left/right)	X,Y,Z
9, 10	Rear Engine support (left/right)	X,Y,Z
11	auxiliary Engine support	X,Y,Z

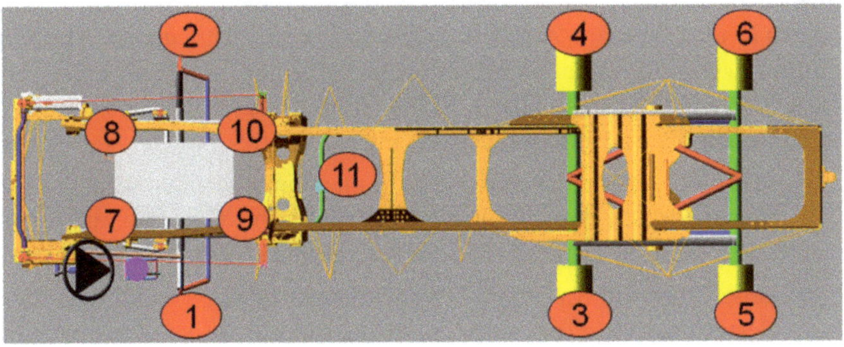

The ability of a vehicle to increase its speed is typically measured in miles per hour per second (mph/s) or kilometers per hour per second (km/h/s).

Acceleration is also connected to vibration analysis, which refers to the rate of change in velocity of an object undergoing vibratory motion. It represents the magnitude and direction of the change in velocity per unit time. When considering acceleration as a vibration function, it typically refers to the variation of acceleration over time in a vibrating system [5].

In a vibrational context, acceleration is measured using accelerometers, which detect and quantify acceleration. These accelerometers provide a time-varying signal representing the acceleration experienced by the analyzed object or system.

The acceleration function describes the behavior of acceleration as a function of time. It provides information about how the acceleration of the vibrating system changes over time, including its magnitude, direction, and any periodic or non-periodic variations. By analyzing the acceleration function, engineers and researchers can gain insights into the dynamics of the vibrating system, including its natural frequencies, resonances, and potential issues such as excessive vibration or structural integrity problems.

For more information, see **J1939-71**, **J1491**, and **J2735** [2, 6, 7].

Acid Trucks (Acidizing Operations)

An acid truck is a specialized commercial vehicle used in the oil and gas industry for acidizing operations. Acidizing is a well-stimulation technique that involves pumping acid or acid-based solutions into a wellbore to enhance the flow of hydrocarbons by dissolving or removing formation materials that might obstruct production.

ADAS (Advanced Driver Assistance System)

ADAS refers to electronic technologies integrated into commercial vehicles to enhance driving safety, convenience, and overall vehicle performance. These systems leverage sensors, cameras, radar, and other technologies to provide real-time data and assistance to drivers, helping them make informed decisions and reducing the risk of accidents. ADAS represents a significant step toward semi-autonomous and autonomous driving capabilities [8].

- ACC: ACC uses radar or cameras to maintain a safe following distance from the vehicle ahead. It automatically adjusts the vehicle's speed to match the traffic flow, reducing the need for constant acceleration and braking.
- Lane Departure Warning (LDW) and Lane Keeping Assist (LKA): LDW alerts the driver if the vehicle unintentionally drifts out of its lane. LKA goes further by gently steering the vehicle back into the lane if the driver does not respond to the warning.
- Forward Collision Warning (FCW) and Automatic Emergency Braking (AEB): FCW detects potential collisions with vehicles or pedestrians in the vehicle's path and alerts the driver. AEB intervenes by automatically applying the brakes if the driver does not respond, mitigating or preventing a collision.
- Blind Spot Monitoring (BSM) and Rear Cross Traffic Alert (RCTA): BSM uses sensors to monitor the vehicle's blind spots and alerts the driver when another vehicle is present. RCTA warns the driver about approaching vehicles while reversing out of a parking spot.

ADB (Air Disc Brakes) (Truck and Bus)

Air disc brakes for commercial vehicles are a highly efficient and modern braking system that utilizes compressed air to apply brake force to the brake disc (rotor). These advanced brakes are commonly employed in heavy-duty commercial vehicles, such as trucks, buses, and trailers, offering enhanced braking performance and safety.

For more information, see **J2115**, **J1587**, **J1860**, and **J680** [1, 9, 10, 11].

AdBlue

AdBlue, refers to a trademarked term for a type of diesel exhaust fluid (DEF). It is a non-toxic, colorless, and odorless fluid used in certain diesel-powered vehicles equipped with selective catalytic reduction (SCR) systems to reduce nitrogen oxide emissions (Figure A.3). AdBlue consists of approximately 32.5% urea and 67.5% deionized water. The urea-selective catalyst reduction system implemented in commercial vehicles facilitates ensuring compliance with the nitrogen oxide (NOx) regulation limit [12].

AdBlue is injected into the exhaust system of a diesel engine equipped with an SCR system, where it reacts with the exhaust gases to produce ammonia. The ammonia then reacts with the NOx in the exhaust to convert it into harmless nitrogen and water vapor [13].

FIGURE A.3 Selective catalytic reduction system using AdBlue or urea fluid [12].

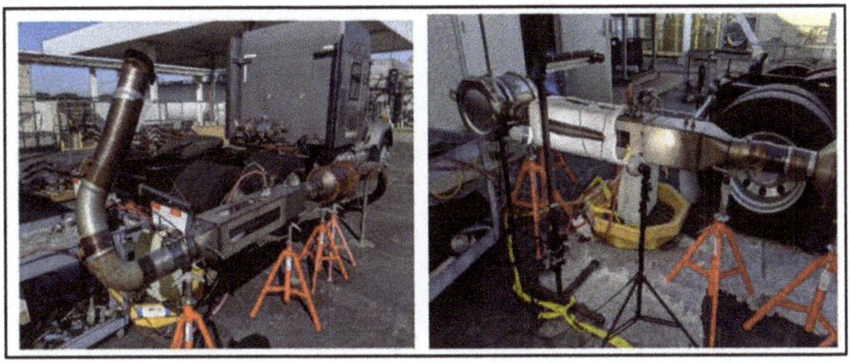

ADEMS (Advanced Diesel Engine Management System)

ADEMS refers to a sophisticated electronic control system that manages and optimizes the operation of a diesel engine in a commercial vehicle. This system integrates various sensors, actuators, and software algorithms to monitor, control, and enhance the diesel engine's performance, efficiency, emissions, and overall operation.

Aerodynamic Drag

Aerodynamic drag is the force that opposes the motion of a vehicle through the air. It is caused by the pressure differential between the front and rear of the vehicle. It can be affected by various factors, such as the shape of the vehicle, the speed of the vehicle, and the properties of the air, such as density and humidity (Figure A.4) [14].

FIGURE A.4 The vehicle shape will impact the aerodynamic drag and fuel efficiency.

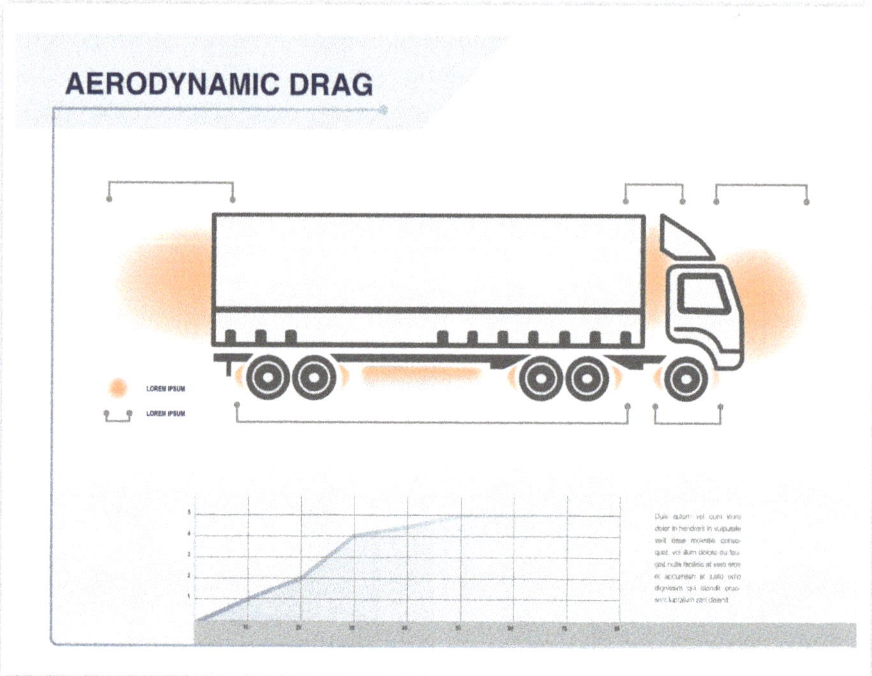

elenabsl/Shutterstock.com.

Aerodynamic drag can be reduced by various means, such as improving the vehicle's aerodynamic shape, reducing its weight, and minimizing air turbulence around it.

For more information, see **J1082**, **J1252**, **J2971**, **J3147**, and **J1270** [14, 15, 16, 17, 18].

Aerodynamics

Aerodynamics studies the motion of air and its interaction with solid objects, such as vehicles. It involves the analysis of various forces acting on the object, such as drag and lift, to determine the most efficient design for a particular application (Figure A.5).

Aerodynamics also includes the study of the flow of gases, such as exhaust gases, through a vehicle's engine and exhaust system. This can help improve engine performance and fuel efficiency.

For more information, see **J1082**, **J1252**, **J2971**, **J3147**, and **J1270** [14, 15, 16, 17, 18].

FIGURE A.5 We use tools like wind tunnels to evaluate the aerodynamics of vehicles.

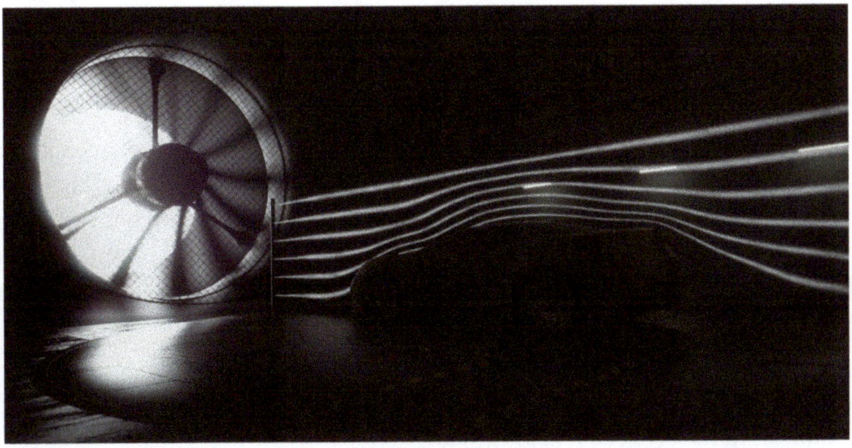

ktsdesign/Shutterstock.com.

Aftercooler (Turbocharger)

An aftercooler refers to a heat exchanger device used in the engine's air intake system to cool down compressed air from a turbocharger or supercharger before entering the combustion chamber. Also known as an intercooler, the aftercooler is crucial in improving engine efficiency, performance, and overall power output [19].

Agricultural Trailer

Agriculture trailers are specialized commercial vehicles transporting various agricultural materials and equipment, and produce between farms, fields, processing facilities, and marketplaces. These trailers are integral to agricultural operations, allowing for efficient movement of goods and resources within the farming industry. Agriculture trailers come in various sizes and configurations to accommodate different types of cargo and equipment.

Agriculture trailers come in several types, including flatbed trailers, dump trailers, livestock trailers, grain trailers, and specialized trailers for transporting specific crops or equipment. The load capacity of agriculture trailers varies based on their design and intended use. They are built to handle various payloads, from bulk crops and livestock to heavy machinery. Agriculture trailers are constructed using durable materials such as steel, aluminum, or both. This ensures that the trailers can withstand the demands of rugged farm environments. Agriculture trailers can be pulled by various vehicles, such as tractors, trucks, or specialized towing equipment. The trailers are equipped with hitching mechanisms compatible with the towing vehicle.

Agriculture trailers are vital in streamlining agricultural logistics, enabling farmers to efficiently transport crops, livestock, machinery, and other essentials within and beyond the farm. Their versatility and specialized designs contribute to productivity and success in modern farming practices.

Agricultural Vehicle

Agricultural vehicles, also known as farm machinery or farm equipment, are specialized vehicles designed and used primarily in the agricultural industry for various farming operations. These vehicles and machines are essential for planting, cultivating, harvesting, and managing crops and livestock. Agricultural vehicles come in varied types and sizes to accommodate different tasks on the farm.

Air ABS

Air ABS is an ABS that uses compressed air to operate the brakes. It is commonly employed in heavy-duty trucks and buses [1].

The system uses sensors to detect when a wheel is about to lock up during braking and then releases and reapplies the brake pressure rapidly to prevent the wheel from locking up. This improves stability and control during braking [20].

For more information, see **J2115**, **J1587**, **J1860**, and **J680** [1, 9, 10, 11].

Air ABS Blink Codes

Air ABS blink codes are diagnostic codes used to troubleshoot issues with an air ABS. When a fault is detected, the system will flash a series of codes on a dashboard warning light to indicate the nature of the problem [21].

Air ABS Brake Modulator

An air ABS brake modulator is an air ABS component that controls the release and application of brake pressure to prevent wheel lockup. The modulator is typically located near the brake valves and regulates the air pressure to the brakes.

The modulator receives signals from the ABS control unit and uses solenoid valves to rapidly release and reapply brake pressure as necessary to prevent wheel lockup.

For more information, see **J1587**, **J1860**, and **J680** [9, 10, 11].

Air ABS Electronic Control Unit (ECU)

An air ABS ECU is the central processing unit of an air ABS. It receives signals from various sensors, such as wheel speed sensors, and uses that information to control the operation of the ABS.

The ECU is responsible for determining when a wheel is about to lock up during braking and sending signals to the brake modulators to release and reapply brake pressure.

For more information, see **J1587**, **J1860**, and **J680** [9, 10, 11].

Air Bag Suspension

An airbag suspension, also known as an air spring bag suspension or airbag suspension, is a type of suspension system used in commercial vehicles to provide a smoother ride, improved load-carrying capacity, and adjustable ride height. Instead of traditional coil or leaf springs, air springs utilize air-filled rubber bags to support the vehicle's weight and absorb road shocks. This technology is found in trucks, buses, trailers, and other heavy-duty vehicles.

Air Brake Compressor

An air brake compressor is a mechanical device that compresses air and delivers it to the air storage tanks in an air brake system. A compressor is used in an air brake system that generates compressed air for use in the brake system. The combined volume of all service and supply reservoirs shall be at least twelve times the combined volume of all service brake chambers at maximum travel of the pistons of diaphragms [22].

For more information, see **J1939**, **J2383**, and **J2384** [2, 23, 24].

Air Brake Dual Air Pressure Gauge

Air brake dual air pressure gauge is a gauge that measures the air pressure in an air brake system's primary and secondary air systems.

Air Brake Reservoir (Air Tank) Drain Valves

Air brake reservoir drain valves, commonly called air tank drain valves, are essential components in commercial vehicles' air brake systems. They play a crucial role in maintaining the safety and performance of the braking system by ensuring the removal of accumulated moisture, oil, and contaminants from the air brake reservoirs (also known as air tanks). This process helps prevent water and debris from compromising the brake system's efficiency and reliability.

Air Brakes (Truck and Bus)

Air brakes, commonly utilized in heavy-duty commercial vehicles such as trucks and buses, are a type of braking system that relies on compressed air to provide efficient and reliable stopping power. This system offers several advantages over-hydraulic brakes, including enhanced safety, handling large vehicles' substantial weight, and stopping demands.

Air Compressor

An air compressor for commercial vehicles is essential to the vehicle's air brake system. It plays a crucial role in generating and maintaining compressed air, which powers the vehicle's air brakes and various other pneumatic systems used in the vehicle's operation. Air compressors are found in various commercial vehicles, including trucks, buses, and heavy machinery.

Air Seeder

An air seeder is a specialized agricultural implement used primarily in modern farming for planting seeds, fertilizing, and applying other inputs (such as herbicides or pesticides) precisely and efficiently. It is essential for large-scale crop cultivation and allows farmers to maximize productivity while minimizing waste.

Air Standard Dual Cycle

The air standard dual cycle is a theoretical thermodynamic model used to analyze and understand the performance of ICEs in commercial vehicles. This model combines elements of the Otto (constant volume) and Diesel (constant pressure) cycles to comprehensively represent the engine's operating characteristics, efficiency, and power output. Analyzing the air standard dual cycle helps engineers and researchers optimize engine design and performance for commercial vehicles.

Air Starter

A commercial vehicle with an air starter is a vehicle or piece of heavy machinery that employs a compressed air starting system to initiate the engine's operation. Instead of relying on a traditional electrical starter motor, these vehicles and machines use compressed air to crank the engine's pistons and start combustion.

Commercial vehicles with air starters are used when traditional electrical starters may not be reliable or suitable, such as in extreme cold weather conditions or environments with the potential for flammable gases or electrical hazards. These air starters are known for their durability, reliability, and ability to provide high torque for starting large diesel engines commonly found in heavy-duty trucks, construction equipment, and industrial machinery.

Air Tank

An air tank, also known as an air reservoir, is an integral component of a commercial vehicle's air brake system. It is fundamental in storing and supplying compressed air to power the vehicle's air brakes and other pneumatic systems. These air tanks are critical for ensuring the safety and reliability of heavy-duty commercial vehicles, including trucks, buses, and trailers.

Regular inspection and maintenance of air tanks are essential to ensure their reliability. This includes checking for air leaks, monitoring the tanks for corrosion, and ensuring proper pressure regulation.

All-Terrain Dump Truck

An all-terrain dump truck refers to a specialized heavy-duty vehicle designed for transporting and unloading materials, such as soil, gravel, sand, and construction debris, in challenging off-road and rugged environments (Figure A.6). This type of dump truck has features that enable it to navigate rough terrains, construction sites, and other demanding locations, making it suitable for various construction, mining, and earthmoving tasks.

FIGURE A.6 An example of an all-terrain dump truck.

Art Konovalov/Shutterstock.com.

Alternator

An alternator is an essential electrical system component in commercial vehicles and other automobiles. It functions as a generator, converting mechanical energy from the vehicle's engine into electrical energy. The alternator is responsible for producing the electricity needed to power the vehicle's electrical systems, charge the battery, and operate various components while the engine runs.

Amboid Gearset

An amboid gearset, or double-cross helical gearset or herringbone gearset, is a specialized gear arrangement used in commercial vehicle transmissions and other machinery. It offers advantages such as smooth operation, reduced noise, and increased load-carrying capacity. The amboid gearset is a helical gear arrangement variation designed to minimize axial forces and improve efficiency.

Ambulance

An ambulance is a specially designed vehicle to transport patients, provide medical care, and offer emergency medical services (Figure A.7). Ambulances are equipped with medical equipment, supplies, and trained personnel to provide medical treatment and transportation to individuals needing urgent medical attention.

For more information, see **J3057**, **J3058**, and **J3027** [25, 26, 27].

FIGURE A.7 Ambulances are medical rooms on wheels.

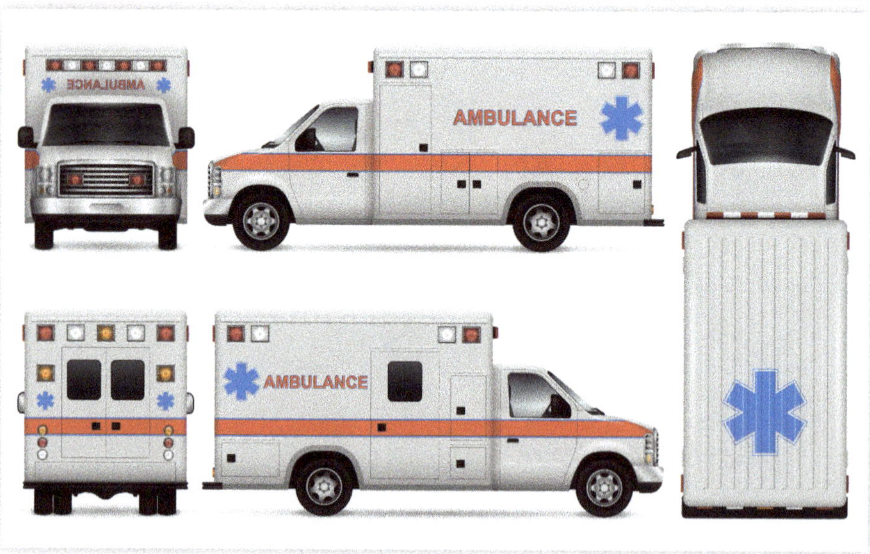

Yuri Schmidt/Shutterstock.com.

AOH (Air-Over-Hydraulic) Brake System

An AOH brake system is a hybrid braking system commonly used in commercial vehicles. It combines elements of both air brakes and hydraulic brakes to provide efficient and reliable braking performance. This system is particularly prevalent in heavy-duty commercial vehicles, such as trucks, buses, and construction equipment.

AOH brake systems include regular inspection of air and hydraulic lines for leaks, monitoring hydraulic fluid levels and quality, and ensuring the integrity of the master cylinder and wheel brake components. Routine checks and servicing are essential for maintaining optimal braking performance.

Armored Truck

An armored truck, also known as an armored vehicle or cash-in-transit (CIT) vehicle, is a specialized commercial vehicle designed to transport valuable cargo, such as currency, precious metals, and useful goods while providing a high level of security against theft, vandalism, and attacks. Armored trucks are crucial in ensuring valuable items' safe and secure movement between banks, businesses, and other locations.

Articulated Bus

An articulated bus refers to a public vehicle designed to carry many passengers. It features a flexible joint, known as an articulation, that allows the bus to bend in the middle, connecting two or more sections. Articulated buses are used in urban and suburban areas to provide efficient and high-capacity transit services. The steering mechanism of an articulated bus is designed to accommodate the bending motion and maintain stability during turns.

Articulated Tractor

An articulated tractor, often referred to as an articulated truck, tractor–trailer, or semi-truck, is a type of commercial vehicle designed for hauling heavy loads over long distances. It consists of two main components: the tractor (front) and the trailer (rear). The tractor is equipped with a powerful engine, driver's cabin, and controls, while the trailer is attached and carries the actual cargo. The connection between the tractor and the trailer is a flexible joint called an "articulation," which allows the two parts to move independently while maintaining a connection.

ASC (Automatic Stability Control)

ASC, often referred to as electronic stability control (ESC) or vehicle stability control (VSC), is an advanced safety feature and technology integrated into commercial vehicles to enhance stability and reduce the risk of skidding or rollovers during challenging driving conditions. ASC is designed to help drivers maintain control of their vehicles, particularly when traction is compromised or the vehicle's stability is at risk.

Aspect Ratio (Tire)

Aspect ratio is a comparison of the tires cross-sectional height to cross-sectional width. These two dimensions are found by measuring the widest point and the tallest point of a fully inflated tire. Aspect ration is determined by the following equation:

$$\text{Aspect ratio} = (\text{section height} / \text{section width}) \cdot 100$$

The ratio of the height of the tire's cross section to its width is expressed as a percentage.

Commercial vehicles' tires are often subject to extreme environments and applications (Figure A.8).

FIGURE A.8 The breakdown of the tire.

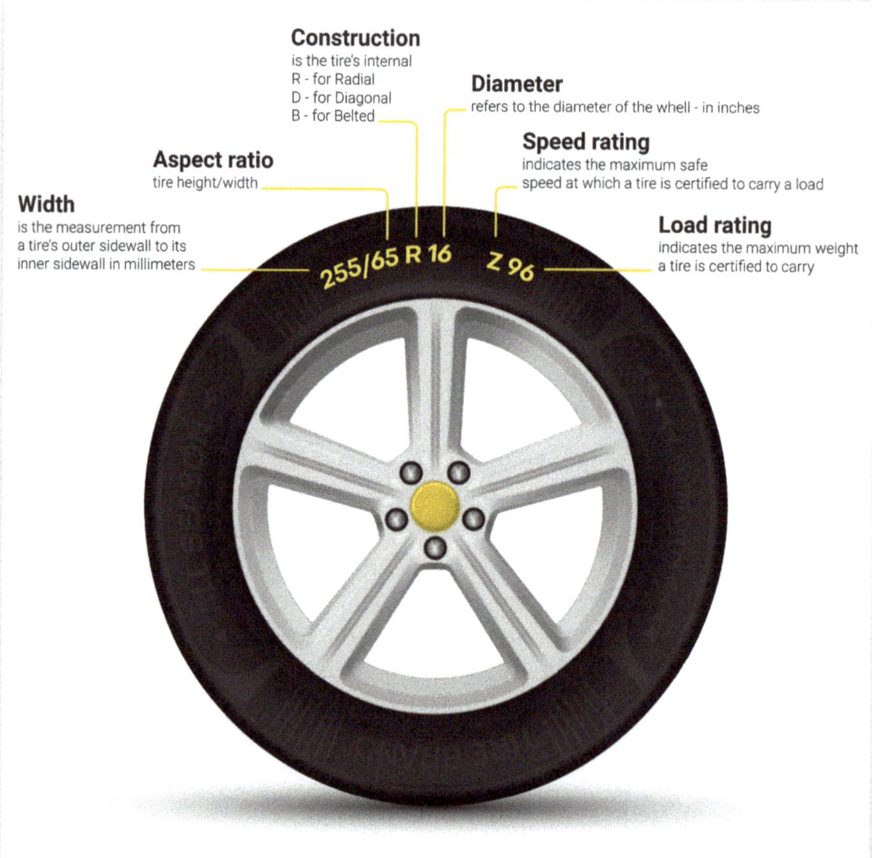

VikiVector/Shutterstock.com.

Asset Tracking

Asset tracking refers to using technology and systems to monitor, manage, and locate valuable assets, such as commercial vehicles, equipment, and inventory, in real time or over time. Asset tracking involves using various technologies, such as Global Positioning System (GPS), radiofrequency identification (RFID), and sensors, to collect and transmit data about the location, status, and movement of assets, providing businesses with enhanced visibility and control over their investments.

ATA (American Trucking Association)

ATA is a prominent trade organization representing the trucking industry's interests in the United States (US). Established in 1933, ATA serves as a leading advocate for the trucking and freight transportation sector, addressing policy issues, promoting safety initiatives, and providing resources for industry professionals. The association is committed to advancing the trucking industry's efficiency, safety, and sustainability through advocacy, education, and collaboration with stakeholders.

ATC (Automatic Traction Control)

ATC is an advanced technology and safety feature designed for heavy trucks and commercial vehicles to improve traction and stability in challenging road conditions. ATC is particularly valuable in situations where some wheels are at risk of losing traction, such as on slippery or uneven surfaces. It operates by intelligently managing the power distribution to the wheels and selectively applying braking to prevent wheel spin and maintain vehicle control.

ATC (Automatic Traction Control) Heavy Truck

ATC is an advanced technology for heavy trucks and other commercial vehicles to optimize traction and stability in challenging road conditions, such as slippery surfaces, uneven terrains, and low-traction situations. ATC enhances the vehicle's performance by adjusting power delivery and braking to individual wheels, thereby preventing wheel spin and improving overall control.

Key features and functions of ATC in a heavy truck context:

- Wheel Slip Detection: ATC utilizes sensors to detect differences in wheel speed, identifying instances of wheel slip or loss of traction. These data are crucial for the system to intervene effectively.
- Engine Power Management: When wheel slip is detected, the ATC system adjusts the engine's power output to prevent excessive throttle application, which could result in wheels spinning and loss of control.
- Brake Application: ATC employs the vehicle's ABS to apply brake pressure to specific wheels experiencing wheel slip. This action redirects power to wheels with better traction, enhancing stability and control.

- Individual Wheel Control: ATC operates on a wheel-by-wheel basis, allowing for precise intervention on wheels experiencing slip while maintaining power and traction on those with better grip.
- Enhanced Stability: By preventing wheel spin and distributing power more effectively, ATC improves the heavy truck's stability and reduces the risk of skidding or jackknifing in challenging road conditions.
- Reduced Tire Wear: Preventing wheel spin through ATC helps extend the lifespan of tires by minimizing excessive wear that can occur when tires lose traction.
- Low-Speed Traction: ATC is especially valuable during low-speed maneuvers, such as climbing steep inclines, navigating icy roads, or driving on loose surfaces, where maintaining traction can be challenging.
- Integration with Other Systems: ATC often works with other vehicle control systems, such as ESC and traction control systems (TCS), to provide comprehensive stability and control management.
- Driver Assistance: While ATC operates automatically, some systems offer driver-selectable modes or overrides to accommodate specific driving preferences or unique situations.
- Safety Enhancement: ATC increases safety for the driver, cargo, and other road users by minimizing the risk of accidents and loss of control due to wheel spin.

Attachments

In commercial vehicles, attachments refer to specialized equipment or devices that can be added or connected to a base vehicle to extend functionality and versatility. These attachments are designed to perform specific tasks, enhance performance, or adapt the vehicle for various applications. Attachments are commonly used in construction, agriculture, logistics, and other industries.

Auto Hauler

An auto hauler, also known as a car carrier or car transporter, is a specialized commercial vehicle designed to transport multiple automobiles from one location to another. Car manufacturers, dealerships, rental companies, and logistics firms commonly use auto haulers to move new or used vehicles efficiently and securely.

Auto haulers are built with multiple levels or decks to accommodate several vehicles simultaneously. This enables transporting a more significant number of cars in a single trip. Auto haulers come in various configurations, including open trailers, enclosed trailers, and semi-trailers, offering options to suit different types of vehicles and transportation needs. Auto haulers are equipped with loading ramps, either hydraulically operated or manually adjusted, to facilitate the safe loading and unloading of vehicles. To prevent movement and damage during transit, auto haulers have securement mechanisms such as straps, chains, wheel chocks, and tie-down points to anchor vehicles in place. Some auto haulers have hydraulic or tilting mechanisms that allow the upper deck to be tilted, making loading and unloading vehicles on the lower deck easier. Auto haulers are designed for efficient vehicle transportation, minimizing the need for individual trips and reducing overall transportation costs. Enclosed auto haulers protect from the elements, decreasing exposure to weather conditions and potential damage during transit.

Automated Transmission

An automated transmission, often called an automated manual transmission (AMT) or automated gearbox, is a sophisticated transmission system used in commercial vehicles, such as trucks, buses, and delivery vans. It combines the convenience of an automatic transmission with the fuel efficiency and control of a manual transmission. An automated transmission is designed to provide a seamless and user-friendly driving experience, making it a popular choice for modern commercial vehicles.

Automatic Gearbox

The automatic gearbox is a type of transmission that uses electronic controls to automatically shift gears based on driving conditions, improving drivability and reducing driver fatigue while enhancing fuel efficiency and reducing emissions (Figure A.9).

FIGURE A.9 An example of the internals of a transmission.

Automatic Hill Start Assist

Automatic hill start assist is an electronic system that automatically holds the brakes on a slope to prevent the vehicle from rolling backward when the driver releases the brake pedal and applies a throttle, improving safety and convenience.

Automatic Transmission

An automatic transmission is a sophisticated system used in various commercial vehicles, such as trucks, buses, and delivery vans. Unlike manual transmissions, which require manual gear shifting, an automatic transmission operates without direct driver intervention. It is an ideal choice for vehicles where driver convenience, efficiency, and ease of operation are essential.

For more information, consult **J3081** [28].

Automatic Transmission Cooler

An automatic transmission cooler is vital to a commercial vehicle's automated transmission system. It dissipates excess heat generated during the transmission's operation, helping maintain the transmission fluid's proper operating temperature. By keeping the fluid within an optimal temperature range, the cooler ensures smooth and efficient transmission performance while prolonging the lifespan of the transmission components.

Autonomous Farm Vehicle

An autonomous farm vehicle is a specialized commercial vehicle designed for agricultural applications that can operate without direct human control. These vehicles use advanced technologies such as GPS, sensors, and artificial intelligence (AI) to perform various tasks on the farm, including planting, harvesting, tilling, and crop monitoring. Autonomous farm vehicles are a critical part of modern agriculture, as they help improve efficiency, reduce labor costs, and enhance overall productivity.

Axle

An axle is a vital component of a commercial vehicle's chassis and drivetrain system, serving as a central structural element responsible for connecting and supporting the vehicle's wheels. It plays a fundamental role in bearing the vehicle's weight, transmitting power from the engine to the wheels, and facilitating steering and suspension systems [29].

Load bearing	Power transmission
Steering and suspension support	Differential integration
Axle type	

For more information, see **J2475** and **J3218** [30, 31].

Axle Back

Axle back refers to a specific configuration or type of exhaust system commonly found in commercial vehicles, particularly trucks and heavy-duty vehicles. This configuration defines the portion of the exhaust system located behind the vehicle's rear axle. The axle back exhaust system optimizes vehicle performance, emission control, and noise reduction.

Axle Configuration

The axle configuration in a commercial vehicle refers to the arrangement of axles and wheels, providing information about the number of axles and the drivetrain configuration (Figure A.10).

The first digit indicates the total number of wheels on the vehicle. The second digit specifies the number of powered (driving) wheels. A few examples are demonstrated below.

- The "4×2" configuration indicates that the vehicle has four-wheel ends, two of which are powered.
- The "6×4" configuration indicates that the vehicle has six-wheel ends, four of which are powered.
- The "6×6" configuration indicates that all six wheels on the vehicle are powered, typically referring to a six-wheel drive (6WD) setup.

FIGURE A.10 Examples of axle configurations; the dark axles are drive axles.

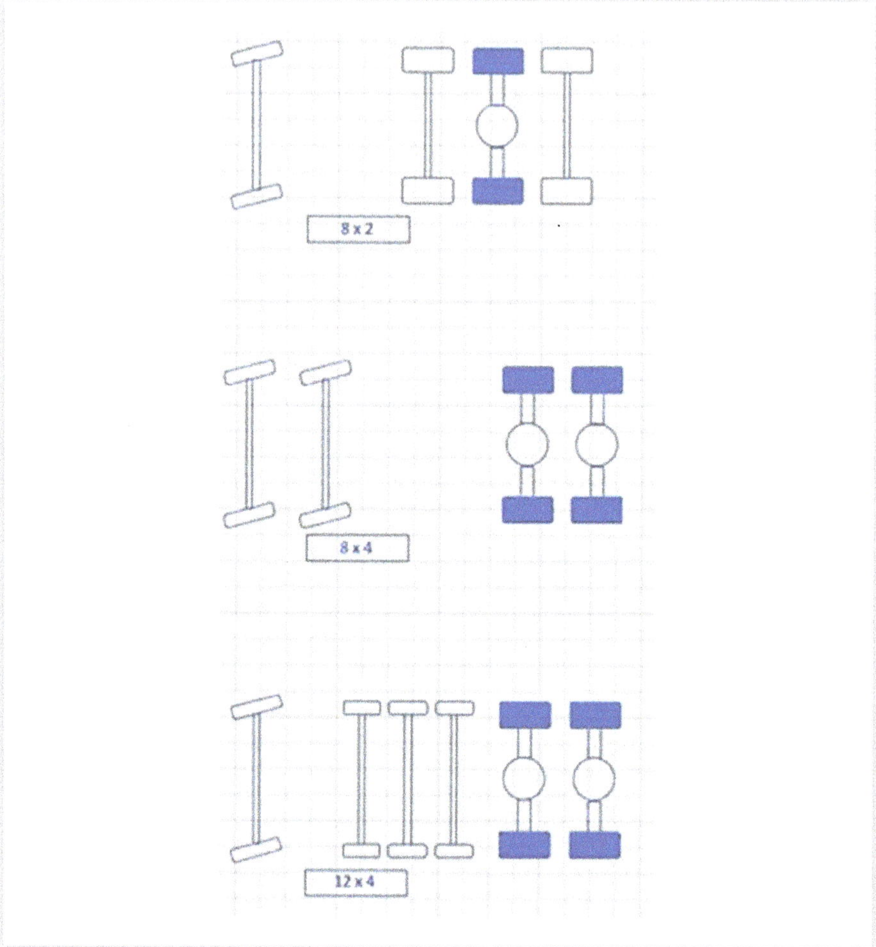

Axle Forward

Axle forward refers to a specific configuration or design characteristic of commercial trucks, particularly those with a cab-over-engine (COE) layout. In this configuration, the front axle of the truck is positioned closer to the front of the vehicle, allowing for maximum use of available cargo space while adhering to legal length restrictions. The axle forward design has distinct advantages in maneuverability, weight distribution, and accommodating larger loads.

Axle Weight

The total weight load that each axle of a vehicle carries is crucial for determining commercial vehicle operations' safety, legality, and efficiency. It includes the weight of the axle itself and the portion of a vehicle's total load borne by the axle. Axle weight is a critical factor in commercial vehicle design and is closely regulated by transportation authorities. Limits are imposed on axle weight to prevent road damage and ensure safe handling characteristics under various load conditions. Exceeding these limits can result in fines and increased maintenance costs due to premature wear on vehicle components and road surfaces. Axle weights are typically measured using weigh stations or portable weigh pads. Regulations vary by jurisdiction but generally set maximum allowable axle weights to balance vehicle capability and infrastructure preservation. Compliance with these regulations is mandatory for commercial vehicle operators [32].

REFERENCES

1. Society of Automotive Engineers, "J2115 Air Brake Performance and Wear Test Code Commercial Vehicle Inertia Dynamometer," SAE Publishing, Warrendale, 2020.
2. Society of Automotive Engineers, "J1939-71 Vehicle Application Layer," SAE Publishing, Warrendale, 2020.
3. Society of Automotive Engineers, "Standard J2627_200908: Braking System," SAE Publishing, Warrendale, 2009.
4. Zhu, H., Yang, Y., Yang, Y., Zeng, J. et al., "Ride Optimization for a Heavy Commercial Vehicle," *SAE Int. J. Commer. Veh.* 7, no. 1 (2014): 150-156, doi:https://doi.org/10.4271/2014-01-0843.
5. Society of Automotive Engineers, "J1013 Measurement of Whole Body Vibration of the Seated Operator of Off-Highway Work Machines," SAE Publishing, Warrendale, 2007.
6. Society of Automotive Engineers, "J1491 Vehicle Acceleration Measurement," SAE Publishing, Warrendale, 2006.
7. Society of Automotive Engineers, "J2735 V2X Communications Message Set Dictionary™ Set," SAE Publishing, Warrendale, 2023.
8. Society of Automotive Engineers, "J3029 Forward Collision Warning and Automatic Emergency Braking Test Procedure and Minimum Performance Requirements - Truck and Bus," SAE Publishing, Warrendale, 2023.

9. Society of Automotive Engineers, "J1587 Electronic Data Interchange Between Microcomputer Systems in Heavy-Duty Vehicle Applications," SAE Publishing, Warrendale, 2013.

10. Society of Automotive Engineers, "J1860 Labeling Air Brake Valves with Their Performance (Input-Output) Characteristics," SAE Publishing, Warrendale, 2016.

11. Society of Automotive Engineers, "J680 Location and Operation of Air Brake Controls in Motor Truck Cabs," SAE Publishing, Warrendale, 2015.

12. Alembath, A., Smith, J.D., Mooney, B., Shinde, A. et al., "Experimental Investigation of Urea Deposit Formation in Selective Catalyst Reduction System to Guide Product Development," *SAE Int. J. Fuels Lubr.* 15, no. 3 (2022): 277-294, doi:https://doi.org/10.4271/04-15-03-0014.

13. IEC, "Diesel Engines—Engine-Mounted Fuel and Urea Tanks and Refuelling Systems for Liquid Urea—Part 1: General Requirements and Test Methods," International Electrotechnical Commission, Geneva, Switzerland, 2006.

14. Society of Automotive Engineers, "J1082 Fuel Economy Measurement Road Test Procedure," SAE Publishing, Warrendale, 2023.

15. Society of Automotive Engineers, "J1252 SAE Wind Tunnel Test Procedure for Trucks and Buses," SAE Publishing, Warrendale, 2012.

16. Society of Automotive Engineers, "J2971 Truck and Bus Aerodynamic Device and Concept Terminology," SAE Publishing, Warrendale, 2019.

17. Society of Automotive Engineers, "J3147 Assessment of Evaluation Method Types to Determine Energy Efficiency of Multi-Vehicle Systems," SAE Publishing, Warrendale, 2022.

18. Society of Automotive Engineers, "J1270 Measurement of Passenger Car, Light Truck, and Highway Truck and Bus Tire Rolling Resistance," SAE Publishing, Warrendale, 2017.

19. Society of Automotive Engineers, "J1994 Laboratory Testing of Vehicle and Industrial Heat Exchangers for Heat Transfer and Pressure Drop Performance," SAE Publishing, Warrendale, 2020.

20. Society of Automotive Engineers, "J2580 Identification and Installation of Air Brake System Components," SAE Publishing, Warrendale, 2022.

21. Society of Automotive Engineers, "J1708 Serial Data Communications Between Microcomputer Systems in Heavy-Duty Vehicle Applications," SAE Publishing, Warrendale, 2016.

22. Society of Automotive Engineers, "J1977 Braking Performance - Rubber-Tired, Self-Propelled Cranes," SAE Publishing, Warrendale, 2019.

23. Society of Automotive Engineers, "J2383 Air Dryer Installation Procedure," SAE Publishing, Warrendale, 2023.

24. Society of Automotive Engineers, "J2384 Air Dryer Test Procedure," SAE Publishing, Warrendale, 2020.

25. Society of Automotive Engineers, "J3057 Ambulance Modular Body Evaluation - Quasi-Static Loading for Type I and Type III Modular Ambulance Bodies," SAE Publishing, Warrendale, 2023.

26. Society of Automotive Engineers, "J3058 Ambulance Interior Storage Compartment Integrity," SAE Publishing, Warrendale, 2022.

27. Society of Automotive Engineers, "J3027 Ambulance Litter Integrity, Retention, and Patient Restraint," SAE Publishing, Warrendale, 2022.

28. Society of Automotive Engineers, "J3081 Heavy Vehicle Controls Prioritization and Conflict Resolution," SAE Publishing, Warrendale, 2020.

29. Society of Automotive Engineers, "J1842 Disc Wheel Hub/Spoke Wheel and Axle Interface Dimensions -Truck and Bus," SAE Publishing, Warrendale, 2023.

30. Society of Automotive Engineers, "J2475 Wheel End Assembly and Axle Spindle Interface Dimensions - Truck and Bus," SAE Publishing, Warrendale, 2020.

31. Society of Automotive Engineers, "J3218 Definition and Measurement of Beam Axle Efficiency," SAE Publishing, Warrendale, 2022.

32. Society of Automotive Engineers, "J2807 Performance Requirements for Determining Tow-Vehicle Gross Combination Weight Rating and Trailer Weight Rating," SAE Publishing, Warrendale, 2020.

B

B20

B20 is a biodiesel fuel blend commonly used in the commercial vehicle industry. It comprises 20% biodiesel and 80% conventional diesel fuel (petrodiesel). B20 is a form of alternative fuel that offers certain environmental benefits and is designed to be compatible with most diesel engines and infrastructure [1].

Biodiesel is a renewable fuel derived from organic materials such as vegetable oils, animal fats, or recycled cooking oil. B20 blends contain a significant proportion of biodiesel, reducing the reliance on fossil fuels and contributing to sustainability efforts. B20 biodiesel blends generally have lower greenhouse gas emissions than conventional diesel fuel. Biodiesel can potentially reduce carbon dioxide, particulate matter, and other harmful emissions, contributing to improved air quality and environmental preservation. B20 is designed to be compatible with most diesel engines without requiring significant modifications. Many modern diesel engines are certified to operate with B20 blends, offering a transition to cleaner fuels without substantial changes.

B20 blends can have performance characteristics similar to conventional diesel fuels in terms of power output, torque, and overall engine operation. However, specific properties of biodiesel, such as its lubricating qualities, can influence engine wear and performance over time. Biodiesel has a higher gel point (the temperature at which it solidifies) than petrodiesel. B20 may require appropriate fuel management in colder climates to prevent fuel gelling and ensure smooth engine operation.

Like any fuel, proper storage and handling practices are essential to prevent contamination and maintain fuel quality. Biodiesel blends can be more sensitive to moisture and microbial growth than conventional diesel fuel. B20 can generally be used with existing diesel fuel infrastructure, such as storage tanks and dispensing equipment. However, it is essential to ensure that the infrastructure is compatible with biodiesel to avoid potential issues.

Back Office

The term back office refers to the administrative and operational functions that support managing, coordinating, and optimizing the use of a fleet of vehicles. It encompasses a range of activities and systems that occur behind the scenes to ensure smooth operations, regulatory compliance, and effective decision-making.

Back office operations often involve specialized fleet management software. This software helps track vehicle locations, monitor driver behavior, manage maintenance schedules, and generate reports for analysis. The back office is crucial in dispatching vehicles to specific locations, optimizing routes, and ensuring timely deliveries. This involves real-time communication with drivers and updating routes based on changing conditions.

Back office functions include driver scheduling, monitoring hours-of-service (HOS) compliance, managing driver assignments, and providing necessary information and support to drivers. The back office tracks vehicle maintenance schedules, schedules repairs, and ensures that vehicles are kept in optimal working condition to minimize downtime.

Back office operations involve staying up-to-date with relevant regulations and ensuring that vehicles and drivers adhere to safety and legal standards. This includes managing licenses, permits, and inspections. The back office collects and analyzes data related to vehicle performance, fuel efficiency, driver behavior, and more. These data are used to make informed decisions for improving fleet operations.

Effective communication between the back office and drivers is essential. This can involve sending instructions, receiving status updates, and addressing any issues or emergencies. Back office systems generate various reports, such as route histories, fuel consumption, maintenance records, and driver performance, which aid in evaluating fleet efficiency and making strategic decisions.

Backhoe

A backhoe refers to a versatile heavy equipment machine used in construction, excavation, digging, and trenching operations. A backhoe consists of a digging arm or boom mounted on the rear of a vehicle, along with a digging bucket attached to the end of the arm. It is commonly employed for various tasks, from digging foundations to clearing debris (Figure B.1).

For more information, see **J326** and **J49** [2, 3].

FIGURE B.1 The backhoe shovel is used to dig trenches.

Ball Joint

A ball joint is critical in commercial vehicles, particularly in the suspension and steering systems. It facilitates the connection between various moving parts, allowing for controlled movement and articulation while maintaining stability and proper alignment. Ball joints are commonly found in the front suspension of vehicles, providing a pivotal connection between the steering knuckle and control arms.

In general, a ball joint consists of a spherical bearing enclosed in a socket, with many variations (Figure B.2). This design allows for rotational movement in multiple directions, making it suitable for articulation and suspension systems [4]. In the suspension system, ball joints connect the vehicle's wheel hub or steering knuckle to the control arms, allowing the wheel to move vertically in response to road irregularities.

FIGURE B.2 Example of some of the ball joint types [4].

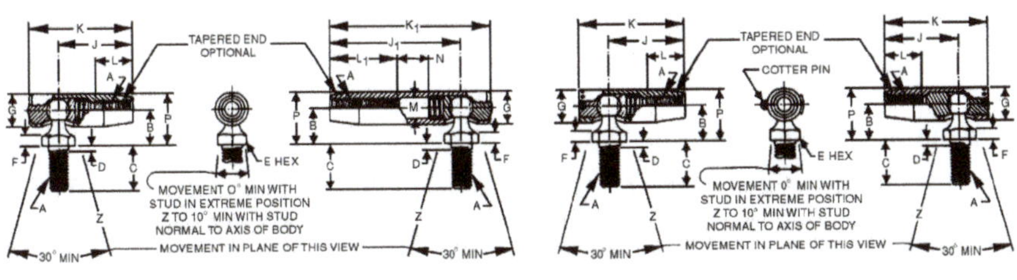

(a) Crimped end plug with spring construction—type A

(b) Crimped end plug with spring construction—type AL

(c) Threaded end plug with spring construction—type B

(d) Threaded end plug without spring construction—type C

Ball joints are also used in the steering system to connect the steering knuckle to the steering linkage, enabling the wheel to turn and change direction. Ball joints help distribute loads from the vehicle's weight and road forces between various suspension and steering components. Ball joints are subject to continuous movement and varying loads, so they are designed with materials and lubrication to withstand wear and friction over the vehicle's lifespan.

Ball joints are essential in off-road vehicles, allowing the suspension to articulate and adapt to uneven terrains while maintaining control. Faulty or worn ball joints can lead to steering instability, uneven tire wear, poor handling, and potential accidents. Regular inspections are essential for identifying issues early. Ball joints are designed to fit specific makes and models of vehicles, ensuring proper fitment and alignment.

Battery

A battery in commercial vehicles refers to an essential electrical component that stores and provides electrical energy to power various systems and components of the vehicle. Batteries are critical in starting the engine, supplying power to lights, electronics, and accessories, and maintaining vehicle functions even when the engine is not running.

Batteries store electrical energy through chemical reactions between their components, typically lead-acid or, more recently, lithium-ion technology. The primary function of a vehicle battery is to provide the initial burst of electrical energy needed to start the engine by turning the starter motor.

Batteries supply electrical power to various vehicle systems, including lights, audio systems, air conditioning, power windows, and electronic components. Commercial vehicles often have auxiliary equipment and accessories that require electrical power, such as refrigeration units, communication systems, and specialized equipment. When the engine is turned off, the battery continues to provide power to essential systems and components, such as lights and security systems. The battery's voltage remains relatively stable, contributing to the overall stability of the vehicle's electrical system.

Cold cranking amps (CCA) measures a battery's ability to deliver high currents in cold temperatures, which is critical for starting the engine in cold climates. The ampere hour (Ah) rating indicates the battery's capacity to have a certain amount of current over a specific duration, which affects how long the battery can power accessories when the engine is off. Lead-acid batteries often require periodic maintenance, such as checking electrolyte levels, ensuring proper terminal connections, and preventing sulfation. The lifespan of a battery depends on factors like usage, temperature, charging habits, and overall maintenance. Batteries generally need replacement every few years.

If the battery is discharged or fails, jumping with another vehicle or an external power source can provide the necessary energy to start the engine.

Lithium-ion batteries are becoming increasingly common in commercial vehicles due to their higher energy density, longer lifespan, and potential for regenerative braking [5].

For more information, see **J537**, **J930**, and **J2185** [6, 7, 8].

Battery Box

A battery box, often called a battery enclosure or battery compartment, is a protective housing designed to contain and safeguard the vehicle's batteries securely. In commercial vehicles, battery boxes are essential to ensure the safe storage, accessibility, and maintenance of the vehicle's battery or batteries.

A battery box provides a dedicated and enclosed space to house the vehicle's batteries, protecting them from external elements, impacts, and environmental conditions. Battery boxes prevent the batteries from moving or shifting during vehicle operation, reducing the risk of damage to the batteries and other vehicle components.

Battery boxes shield batteries from moisture, chemicals, and corrosive substances, extending the lifespan of the batteries and maintaining their performance. Battery boxes are designed with hinged lids, access panels, or removable covers to allow easy access for maintenance, inspections, and battery replacements. Some battery boxes incorporate ventilation systems to dissipate heat and prevent the accumulation of potentially explosive gases produced during battery charging.

Battery boxes include insulation and grounding features to prevent accidental contact with electrical components, reducing the risk of electrical shocks and hazards. Battery boxes are typically designed for secure attachment to the vehicle's chassis, ensuring stability and preventing movement during transit.

Battery Disconnect

A battery disconnect, a battery cutoff switch or master switch, is an electrical component used in commercial vehicles to disconnect the vehicle's battery from the electrical system. It is a safety and maintenance feature, allowing the vehicle's electrical power to be completely cut off when needed, thereby preventing battery drain and electrical malfunctions and enhancing safety during maintenance or storage.

A battery disconnect switch physically interrupts the electrical connection between the vehicle's battery and the rest of the electrical system, preventing current flow. In electrical or engine-related emergencies, activating the battery disconnect can quickly halt all power to the vehicle, minimizing the risk of fires, short circuits, or other electrical issues. When a vehicle is not in use for an extended period, a battery disconnect helps prevent gradual battery discharge due to the continuous draw of electrical components in standby mode.

Before performing maintenance or repairs on a commercial vehicle, activating the battery disconnect ensures no power is supplied to the vehicle's systems, reducing the risk of accidental electrical shocks or hazards. By disconnecting the battery, a battery cutoff switch can deter potential thieves from starting the vehicle since it effectively immobilizes the engine and prevents ignition. During long periods of vehicle storage, such as in seasonal industries, a battery disconnect prevents battery drain and helps maintain the battery's charge.

Battery Isolation Systems

Battery isolation systems, also known as battery isolators or battery disconnect systems, are electrical components or devices used in commercial vehicles to control the electrical connection between the vehicle's primary battery and auxiliary batteries or systems. These systems play a crucial role in managing power distribution, preventing battery drain, and ensuring the availability of electrical power when needed.

Battery isolation systems allow the driver or operator to control the flow of electrical power between the primary battery and auxiliary batteries or systems. Many commercial vehicles have auxiliary batteries that power accessories like lights, refrigeration units, communication equipment, or other specialized systems.

Battery isolators help prevent situations where auxiliary equipment or accessories drain power from the primary battery, ensuring that the vehicle's engine can always start. Vehicles with multiple batteries, such as dual-battery setups, benefit from isolation systems to manage power distribution effectively. Battery isolators can operate automatically, disconnecting auxiliary batteries when the engine is off or manually, allowing the driver to control the connection.

Some isolation systems use voltage sensing technology to monitor the voltage levels of the primary and auxiliary batteries and automatically disconnect or connect them as needed. Isolation systems often utilize solenoids or relays that open or close the electrical circuit between batteries based on the driver's command or predetermined conditions.

Battery isolators sometimes prioritize charging the primary battery before diverting power to auxiliary batteries, ensuring that the vehicle's main starting battery remains fully charged. Battery isolation systems require that auxiliary batteries stay charged and operational even if the primary battery is drained or disconnected. Maintaining the charge of the primary battery is essential for vehicle safety systems, such as airbags, ABSs, and other electronic components.

Some battery isolation systems allow customization based on the vehicle's electrical requirements and specific usage scenarios. In emergency vehicles or work trucks with specialized equipment, battery isolation systems can ensure power availability for critical functions.

Battery State-of-Charge (SOC) Indicator

A battery SOC indicator, commonly called a battery gauge or battery meter, is a visual or digital display on commercial vehicles that provides information about the current charge level in the vehicle's battery. This indicator helps drivers and operators monitor the battery's energy status, allowing them to make informed decisions about electrical usage and potential recharging needs.

The battery SOC indicator displays the remaining electrical energy in the battery as a percentage, clearly showing how much charge is available. Battery SOC indicators can be visual displays with segments, bars, or light-emitting diodes (LEDs) that light up to indicate charge levels or digital screens showing exact percentages. The indicator provides real-time information about the battery's current charge level, allowing drivers to gauge whether the battery has sufficient energy for intended operations. By monitoring the battery's SOC, drivers can manage electrical loads and avoid depleting the battery to a level that might compromise starting the engine.

Battery SOC indicators help drivers avoid excessive battery discharge, ensuring enough charge is retained to start the vehicle. The indicator assists drivers in determining when to recharge the battery, preventing situations where the battery is drained to a critical level.

Battery SOC indicators contribute to driver awareness of energy consumption and electrical system status, promoting responsible usage. In some vehicles, the battery SOC indicator can reflect the alternator's performance by showing how effectively it charges the battery. For vehicles with auxiliary equipment, such as refrigeration units or communication systems, the indicator helps drivers monitor energy availability.

In vehicles with multiple batteries, the indicator may show the SOC for each battery separately. In hybrid or electric vehicles (EVs), the battery SOC indicator may change as regenerative braking generates electricity, recharging the battery. Some indicators include low battery warning lights or alerts when the battery's charge drops to a critical level.

Battery System

A battery system refers to the integrated network of components responsible for storing, distributing, and supplying electrical energy to power various vehicle systems, accessories, and components. The battery system is crucial in starting the engine, providing power to lighting, electronics, and auxiliary equipment, and maintaining vehicle functions even when the engine is not running.

Battery Terminal

A battery terminal, often called a terminal, is a metal connection point on a commercial vehicle's battery that serves as the interface between the battery and the vehicle's electrical system. Battery terminals play a crucial role in facilitating the transfer of electrical energy from the battery to various vehicle components and accessories.

BBC (Bumper Back of Cab)

Distance BBC is a critical measurement in the design and classification of commercial vehicles, particularly in the trucking industry (Figure B.3). BBC refers to the distance from the front bumper of a commercial vehicle to the rear-facing surface of the cab or driver compartment. This measurement plays a significant role in determining the overall length and maneuverability of the vehicle.

BBC is measured in linear units, such as inches or millimeters, and represents the length between the front bumper and the back of the cab or the front surface of the driver compartment. BBC is a key factor in calculating the overall length of the vehicle, which is often regulated by transportation authorities and affects matters like vehicle registration, licensing, and access to certain roadways. The BBC measurement influences the vehicle's maneuverability, including its ability to make turns, navigate tight spaces, and operate safely on different types of roads.

Different regions and governing bodies may have regulations and classifications based on BBC measurements. For example, certain classes of trucks may have specific BBC requirements for safety and operational reasons. The BBC can vary significantly depending on the type and purpose of the commercial vehicle. For example, long-haul semi-trucks often have longer BBC measurements to accommodate sleeper cabs, while shorter BBCs are common in urban delivery trucks for maneuverability. In longer BBC configurations, there is often more space available for driver amenities, such as sleeping quarters, seating arrangements, and storage compartments.

The BBC can also affect the cargo capacity of a vehicle, as it may determine the available space for attaching trailers, containers, or additional equipment.

FIGURE B.3 Various dimensions on a commercial truck; note the BBC [9].

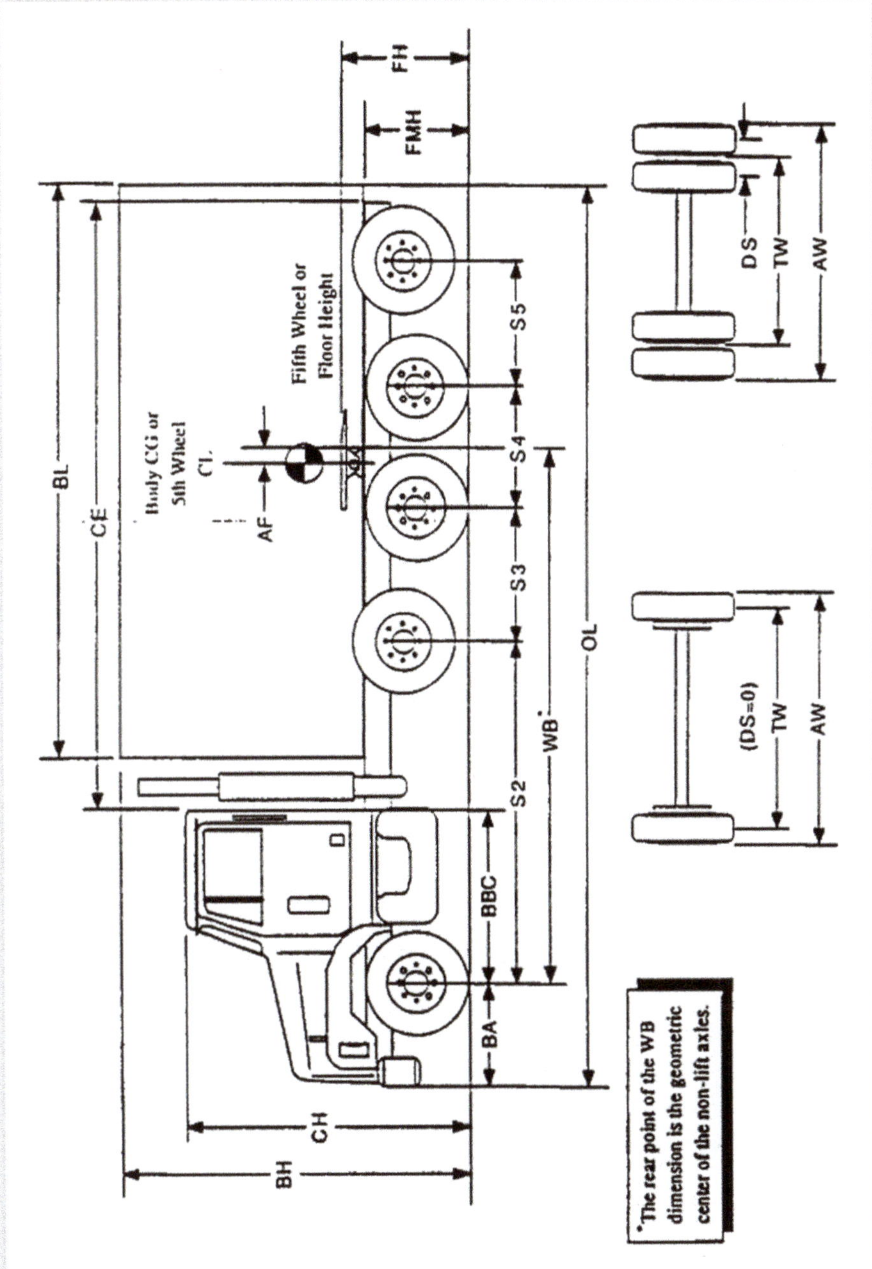

Beam Axle

A beam, solid, or live axle is a type of suspension system used in commercial vehicles. It consists of a single rigid axle that connects the wheels on either side of the vehicle. Unlike independent suspension systems, where each wheel can move independently, a beam axle allows both wheels on the same axle to move together as a single unit.

Belly Dump Trailer

A belly dump trailer, also known as a bottom dump trailer or hopper trailer, is a specialized type of commercial trailer designed for efficiently transporting and unloading bulk materials, such as aggregates, gravel, sand, and other loose substances. Its unique design allows the cargo to be offloaded from the bottom of the trailer, making it well-suited for construction and road-building projects.

A belly dump trailer features a hopper-shaped container with a bottom opening. The hopper design allows for controlled unloading of bulk materials. The distinctive feature of a belly dump trailer is its ability to unload the cargo by opening the bottom of the hopper. This unloading method facilitates the spreading of materials along the intended path. Bottom unloading enables faster and more precise unloading compared to traditional dump trailers, minimizing the need for additional equipment to spread the material.

Operators can control the material discharge rate by adjusting the hopper's bottom opening, allowing for controlled distribution of materials during unloading. Belly dump trailers often result in cleaner unloading than end dump trailers, reducing spillage and minimizing environmental impact. To unload, the trailer is driven forward, while the bottom of the hopper is opened, allowing the material to flow out. This method contributes to smoother unloading and safer operation. Belly dump trailers come in various sizes and capacities, catering to hauling requirements and payload weights.

Belly dump trailers are commonly used in construction and road-building projects to transport and distribute gravel, aggregates, and asphalt. The efficient unloading and controlled spreading make belly dump trailers suitable for projects in off-road or challenging terrains. Belly dump trailers are often equipped with tandem or tri-axle configurations to support the weight of the loaded cargo and ensure proper weight distribution. The bottom opening of the hopper is operated by a hydraulic system that allows for precise control over the unloading process.

Belt Tensioner

A belt tensioner is a mechanical component used in commercial vehicles to maintain the proper tension of belts that drive various engine accessories, such as the alternator, power steering pump, air conditioning compressor, and water pump.

It ensures optimal belt performance by keeping the belt at the correct tension, preventing slipping, and ensuring efficient power transmission.

BHP (Brake Horsepower)

BHP measures an engine's power output when it exits the engine, before any power losses due to the drivetrain and other components. It represents the power the engine produces that is available to perform work, such as moving the vehicle.

Biodiesel

Biodiesel, a type of alternative fuel, is a renewable and environmentally friendly diesel fuel substitute derived from biological sources such as vegetable oils, animal fats, and used cooking oils. It is designed to be compatible with traditional diesel engines, offering a more sustainable and cleaner-burning option for commercial vehicles.

For more information, see *Impacts of Biodiesel on the Durability of an Advanced After-Treatment Diesel Engine* [10].

Block Heater

A block heater, also known as an engine block heater or coolant heater, is used in commercial vehicles to preheat the engine's coolant or the engine block itself, especially in cold climates. It helps improve engine startup performance, reduce wear and tear, and lower emissions by maintaining a warmer operating temperature during cold weather conditions.

Body Builders

Body builders are specialized manufacturers or companies that design, engineer, and construct custom bodies and structures mounted onto commercial vehicles' chassis. These custom bodies can serve various purposes and are tailored to the specific needs of businesses and industries, including cargo transport, service vehicles, and emergency response units.

Boom Lift Truck

A boom lift truck (also known as boom truck), cherry picker, aerial lift truck, or bucket truck is a specialized commercial vehicle designed for accessing elevated areas and performing tasks requiring height elevation (Figure B.4). This versatile vehicle combines a truck chassis with an extendable boom or aerial platform, providing workers with a safe and stable elevated workspace for maintenance, construction, utility work, and emergency services.

FIGURE B.4 An example of a boom lift truck.

Tricky_Shark/Shutterstock.com.

Boundary Mapping

Boundary mapping is a technique that uses GPS to define and map the boundaries of fields for accurate planting and management.

Box Truck

A box truck, also known as a cube truck, van body truck, or box van, is a versatile commercial vehicle characterized by its enclosed cargo compartment resembling a rectangular box. Designed primarily for transporting goods, the box truck offers a secure and weatherproof environment for cargo, making it a popular choice for various industries that require efficient and protected transportation.

Box Van

A box van, often called a cargo van or enclosed van, is a specialized commercial vehicle designed for transporting goods and cargo within an enclosed and weatherproof cargo compartment (Figure B.5). Distinctive for its compact design and secure storage space, the box van is a versatile choice for businesses that require efficient and protected transportation of items, making it an essential component of the logistics and delivery industry.

FIGURE B.5 Inner city delivery is easier accommodated with smaller vehicles such as box vans.

Gorodenkoff/Shutterstock.com.

Brake Application Air

Brake application air, often referred to simply as brake air, is a vital component of the air brake system in commercial vehicles, particularly in trucks and buses. This system uses compressed air to apply and release the brakes, ensuring safe and reliable stopping power. Brake application air plays a central role in controlling and maintaining the vehicle's braking function.

For more information, see **FMVSS 121** [11].

BSD (Blind Spot Detection)

BSD is an advanced safety technology integrated into commercial vehicles to enhance driver awareness and reduce the risk of accidents caused by vehicles or objects in the driver's blind spots. It utilizes sensors and warning systems to alert the driver when another vehicle or obstacle is present in areas not easily visible through standard mirrors.

For more information, see *Unsettled Topics in Obstacle Detection for Autonomous Agricultural Vehicles* [12].

Bucket Truck

A bucket truck, a cherry picker, an aerial lift truck, or an elevated work platform is a specialized commercial vehicle designed to provide formal access for workers to perform tasks at height. This versatile vehicle combines a truck chassis, an extendable boom, and an elevated platform or bucket, offering a safe and stable workspace for maintenance, construction, utility work, and emergency services.

Bulkhead

A bulkhead refers to a structural partition or barrier separating the cab area from the cargo area in certain types of vehicles, such as vans, trucks, and buses. Also known as a divider or partition wall, the bulkhead serves various purposes, including safety, security, noise reduction, and organization within the vehicle's interior.

Bulldozer

A bulldozer is a powerful, heavy-duty commercial vehicle commonly used in construction, mining, and earthmoving applications. It is characterized by its large, broad blade at the front, known as a "dozer blade" or "blade," and a robust crawler undercarriage (Figure B.6). Bulldozers are designed for tasks such as pushing, excavating, grading, and leveling terrain with exceptional strength and precision.

FIGURE B.6 The dominant feature of the bulldozer is the front blade.

A bulldozer is a robust and versatile commercial vehicle used for earthmoving, land clearing, excavation, grading, and leveling in various industries. Its robust capabilities and rugged design make it an essential machine in construction, mining, agriculture, and forestry operations.

Bus

A bus is a commercial vehicle designed to transport passengers, typically in large numbers, over fixed routes or for various purposes such as public transportation, tourism, education, and private charter services. Buses come in multiple sizes, configurations, and types to meet the diverse transportation needs of communities and businesses (Figure B.7).

FIGURE B.7 An example of a commuter bus.

Martinez Alonso/Shutterstock.com.

REFERENCES

1. Pillai, R., Brusstar, M., Boehman, A., and Ludlam, S., "Evaluating the Performance of a Conventional and Hybrid Bus Operating on Diesel and B20 Fuel for Emissions and Fuel Economy," SAE Technical Paper 2020-01-1351 (2020), doi:https://doi.org/10.4271/2020-01-1351.
2. Society of Automotive Engineers, "J326 Nomenclature - Hydraulic Backhoes," SAE Publishing, Warrendale, 2020.
3. Society of Automotive Engineers, "J49 Specification Definitions - Hydraulic Backhoes," SAE Publishing, Warrendale, 2020.
4. Society of Automotive Engineers, "J490 Ball Joints," SAE Publishing, Warrendale, 2012.
5. Kolodziejczyk, B., "Emerging Automotive Battery Chemistries: Hedging Market Bets," SAE Publishing, Warrendale, 2023, doi:https://doi.org/10.4271/EPR2023008.
6. Society of Automotive Engineers, "J537 Storage Batteries," SAE Publishing, Warrendale, 2023.
7. Society of Automotive Engineers, "J930 Storage Batteries for Off-Road Self-Propelled Work Machines," SAE Publishing, Warrendale, 2022.
8. Society of Automotive Engineers, "J2185 Life Test for Heavy-Duty Storage Batteries (Lead Acid Type Only)," SAE Publishing, Warrendale, 2018.
9. Society of Automotive Engineers, "J1574 Measurement of Vehicle and Suspension Parameters for Directional Control Studies," SAE Publishing, Warrendale, 2018.
10. Ku, Y., Lu, J., and Lin, K., "Impacts of Biodiesel on the Durability of an Advanced After-Treatment Diesel Engine," *SAE Int. J. Fuels Lubr.* 10, no. 1 (2017): 1-9, doi:https://doi.org/10.4271/2016-01-9080.
11. US Department of Transportation, "Federal Motor Vehicle Safety Standard 121 Warning Signals, Air Pressure and Vacuum Gauges," National Highway Traffic Safety Administration, Washington DC, 2005.
12. Moorehead, S., "Unsettled Topics in Obstacle Detection for Autonomous Agricultural Vehicles," SAE Publishing, Warrendale, 2021, doi:https://doi.org/10.4271/EPR2021029.

C

Cab

The cab, short for cabin or driver's cab, is the enclosed space in a commercial vehicle where the driver and often a co-driver or passengers sit (Figure C.1). It is the front section of the vehicle's chassis. It serves as the control center for the vehicle, housing the driver's seat, controls, and various amenities necessary for comfortable and safe operation.

FIGURE C.1 The driver spends considerable time in the truck cab.

Scharfsinn/Shutterstock.com.

Cab Forward

Cab forward, often referred to as forward control or front-engine design, is a vehicle configuration commonly used in commercial vehicles where the driver's cabin (cab) is positioned forward of the vehicle's front axle. This design optimizes available space, enhances visibility, and provides specific advantages in various industries and applications.

Car Hauler Trailer

A car hauler trailer, also known as an auto-transporter or car carrier trailer, is a specialized type of commercial vehicle trailer designed to simultaneously transport multiple automobiles, vehicles, or light trucks (Figure C.2). These trailers are widely used in the automotive industry, vehicle transportation services, and car dealerships to efficiently transport vehicles over medium to long distances.

FIGURE C.2 Specialized vehicles that haul cars.

Vitpho/Shutterstock.com.

Cargo Box

A cargo box, bed, cargo compartment, or area is an essential feature of various commercial vehicles designed to transport goods, equipment, and materials. This enclosed or open space serves as the designated area for securely storing and transporting cargo, making it a fundamental component in logistics and transportation.

Cargo Van

A cargo van, also known as a panel van or delivery van, is a versatile type of commercial vehicle designed primarily for transporting goods, packages, and cargo. With its enclosed cargo compartment and practical design, the cargo van serves as an essential tool in the transportation, logistics, and delivery industries.

CBE (Cab Beside Engine)

CBE, also known as cab beside or side-mounted cab, refers to a commercial vehicle configuration in which the driver's cabin (cab) is positioned adjacent to the engine compartment rather than directly over it. This design allows for unique advantages in terms of space utilization, visibility, and vehicle dynamics, particularly in specific applications and industries.

CCA (Cold Cranking Amps)

A CCA battery is a specialized type of lead-acid battery used in commercial vehicles to provide the high electrical current required to start the vehicle's engine, especially in cold weather conditions. CCA is a crucial specification for determining a battery's ability to start the engine in cold temperatures, making it a key consideration in the selection of batteries for commercial vehicles.

For more information, see **J2185** [1].

Cement Truck (Well Cementing Operations)

A cementing truck refers to a specialized commercial vehicle used in the oil and gas industry for well cementing operations. Cementing trucks are equipped with the equipment to mix cement slurries and pump them into wellbores to provide zonal isolation, secure casing, and support integrity.

Center Differential

A center differential, often referred to simply as a "center diff," is a crucial component found in many four-wheel drive (4WD) and all-wheel drive (AWD) commercial vehicles, particularly those designed for off-road, rugged, or all-terrain use. This differential is situated within the vehicle's drivetrain and plays a vital role in distributing power between the front and rear axles, improving traction and maneuverability.

CFE (Cab Forward Engine)

See **Cab Over Engine**.

Charge Air Cooler (CAC)

CAC is critical in commercial vehicles' engine cooling and air intake systems equipped with turbochargers or superchargers. Also known as an intercooler, it plays a crucial role in improving engine efficiency and performance by cooling the compressed air before it enters the engine's combustion chambers.

Chassis

The chassis serves as the foundational structure or framework upon which the various components and systems of the vehicle are mounted or integrated. It forms the structural skeleton of the vehicle, providing support and alignment for the engine, transmission, suspension, body, and other essential components. The term "chassis" is used broadly to describe the underlying structure of various commercial vehicles, including trucks, buses, and specialized vehicles like ambulances and delivery vans.

City Bus

City buses refer to large motor vehicles designed to transport passengers within urban areas and cities. City buses are crucial to public transportation systems, providing convenient and efficient transit options for residents and commuters.

Class 4

Class 4 vehicles, as categorized by vehicle classification systems, represent a specific segment of commercial vehicles based on their gross vehicle weight rating (GVWR) and intended use. These vehicles typically fall within a certain weight range and are subject to specific regulations and requirements based on their classification. Class 4 vehicles encompass a variety of truck types designed to fulfill a range of transportation needs across different industries.

Class 5

Class 5 vehicles, as defined by vehicle classification systems, represent a distinct segment of commercial vehicles based on their GVWR and intended use. Class 5 vehicles offer enhanced payload capacities and are designed to tackle a range of applications across various industries.

Class 6

Class 6 vehicles, as defined within vehicle classification systems, constitute a specific category of commercial vehicles based on their GVWR and intended use. Besides a particular range of operating weight, class 6 vehicles are characterized by their capability to handle substantial payloads while offering versatility across various industries and applications.

Class 7

Class 7 vehicles, as defined by vehicle classification systems, constitute a distinct category of commercial vehicles based on their GVWR and intended use. Besides a particular range of operating weight, class 7 vehicles are characterized by their ability to handle substantial payloads and provide versatile solutions for various industries and applications.

Class 8

Class 8 vehicles, defined within vehicle classification systems, form a distinct and significant category of commercial vehicles based on their GVWR and intended use. Besides a particular range of operating weight, class 8 vehicles are characterized by their heavy-duty capabilities and versatility, serving as the backbone of various industries and transportation sectors.

CNG (Compressed Natural Gas)

CNG is a clean and alternative fuel that primarily consists of methane compressed to a higher pressure for use as a fuel in vehicles. Standards and terminology ensure clarity and consistency within the automotive industry, particularly concerning the design, performance, and safety aspects of vehicles utilizing CNG.

Coach Buses

Coach buses refer to large, comfortable, and luxurious motor vehicles designed for long-distance travel, sightseeing tours, and intercity transportation. Coach buses are known for their spacious interiors, amenities, and comfortable seating, making them popular for group travel and tourism.

COE (Cab Over Engine)

COE refers to a specific type of commercial vehicle design in which the driver's cabin (cab) is positioned directly above the engine compartment. Also known as cab forward or cab forward engine, this configuration maximizes cargo space, maneuverability, and visibility, making it a popular choice for various industries and applications.

Coffee Harvester

A coffee harvester is a specialized commercial vehicle or agricultural machine designed for the efficient and labor-saving harvesting of coffee beans from coffee plants. These machines play a crucial role in the coffee industry, automating the picking of coffee cherries containing valuable coffee beans.

Coffee harvesters are purpose-built machines designed to navigate coffee plantations and effectively harvest coffee cherries while minimizing plant damage. Coffee harvesters use a variety of mechanisms to harvest coffee cherries, including vibration, mechanical shakers, or rotating brushes. These mechanisms dislodge ripe cherries from the coffee bushes. Many coffee harvesters have adjustable settings to accommodate different coffee plant varieties and ripeness levels, ensuring an efficient harvest. Harvested coffee cherries are collected in bins, hoppers, or conveyor systems on the harvester for later processing and sorting. Coffee cherries are delicate, and coffee harvesters are designed to handle them gently to prevent damage, ensuring the quality of the coffee beans.

Coffee plantations often have uneven terrain and steep slopes. Coffee harvesters are equipped to navigate such terrain while maintaining stability. The capacity of a coffee harvester can vary from small, manually operated machines for small-scale farms to more extensive, motorized harvesters for large commercial plantations. Coffee harvesters significantly reduce the labor required for coffee harvesting, making the process faster and more cost-effective.

Cold Chain Management

For industries that require temperature-controlled transportation, cold chain management systems monitor and maintain temperature-sensitive cargo throughout the supply chain.

Combine Harvester

A combine harvester, often, harvests and cleans cereal crops such as wheat, barley, oats, and corn. Combines are indispensable in modern agriculture as they significantly increase the efficiency of crop harvesting and reduce the labor required.

Combines are equipped with a cutting head or header at the front that gathers and cuts standing crops (Figure C.3). These crops are then conveyed into the machine, where the threshing mechanism separates the grain (e.g., wheat kernels) from the plant (straw or stalk). The separated grain goes through a cleaning system that removes any remaining chaff, dust, and impurities, ensuring that only clean grain is collected in the grain tank. Combines typically have a large grain tank or hopper that stores the harvested grain until it can be unloaded into a waiting grain cart or truck. Different headers are available to suit specific crops and conditions, including headers for small grains, corn, soybeans, and sunflowers.

FIGURE C.3 The combine gathers from the field and moves the harvest to the waiting trailer.

bbernard/Shutterstock.com.

Combines have adjustable settings to optimize harvesting for different crops, crop conditions, and yield levels. Most modern combines feature an enclosed operator cab with ergonomic controls, air conditioning, and modern technology, enhancing operator comfort and productivity. Combines are equipped with an unloading auger or conveyor system that transfers the harvested grain from the grain tank into a waiting transport vehicle. Many modern combines are equipped with GPS and yield monitoring technology that provides real-time data on crop yield and performance in the field.

Common Rail Fuel Injection

Common rail fuel injection is an advanced and efficient fuel delivery system used in many modern commercial vehicles, particularly diesel engines. It revolutionized the way fuel is injected into the engine cylinders by providing precise control over the injection process, resulting in improved fuel efficiency, reduced emissions, and enhanced engine performance.

Compact Tractor

A compact tractor is a versatile and relatively small agricultural or utility tractor designed for various tasks on farms, estates, and other properties. These tractors are known for their maneuverability, ease of operation, and compatibility with various attachments and implements. Compact tractors are widely used in landscaping, gardening, small-scale farming, and property maintenance.

Concrete Mixer

A concrete mixer is a versatile and essential piece of equipment used in construction and masonry applications to efficiently combine cement, aggregates (such as sand and gravel), water, and sometimes additives into a homogeneous concrete mixture. To accommodate different project requirements, these mixers are available in various sizes and configurations, including portable, stationary, and truck-mounted models.

Concrete Mixer Truck

A concrete mixer truck, often called a cement mixer truck or simply a mixer truck, is a specialized commercial vehicle designed to transport ready-mixed concrete to construction sites. These vehicles are equipped with a rotating drum or mixer drum that continuously agitates the concrete mixture during transit to maintain its consistency and quality until it is ready to be poured.

The most distinctive feature of a concrete mixer truck is the large cylindrical drum located on the rear of the vehicle. This drum is equipped with rotating blades or fins that continuously mix the concrete to prevent it from settling or hardening during transportation. Mixer trucks come in various sizes, with drum capacities ranging from a few cubic yards to several cubic yards of concrete. The size and requirements of the construction project determine the capacity. Ready-mixed concrete is loaded into the drum at a concrete batching plant or mixing site. The mixer truck's drum rotates in transit to keep the concrete mixture homogeneous and workable.

Operators have control over the speed and direction of drum rotation to adjust the mixing process according to the specific requirements of the concrete mix. Mixer trucks use a hydraulic system to power the rotation of the drum and control its discharge. This system ensures consistent and reliable mixing and unloading. At the construction site, the concrete can be discharged through a rear chute, typically extending from the rear of the drum. Some mixer trucks may also have side chutes or conveyor belts for more precise placement of the concrete. The constant agitation of the concrete within the mixer drum prevents segregation of the materials, ensuring a uniform mix and maintaining the concrete's workability.

Conestoga Trailer

A Conestoga trailer, often called a Conestoga, is a type of semi-trailer used in the transportation industry, particularly for hauling freight that requires protection from the elements and secure loading and unloading. What sets the Conestoga trailer apart is its unique sliding tarp system, which provides quick and easy access to the cargo area without needing traditional manual tarping and untarping procedures.

Conspicuity Lighting

Conspicuity lighting refers to a system of lights and reflective materials strategically placed on commercial vehicles to enhance their visibility, especially during low-light conditions or adverse weather [2]. Conspicuity lighting is designed to reduce the likelihood of accidents by making the vehicle more noticeable to other drivers and pedestrians, ultimately improving road safety.

For more information, see **J1967**, **J759**, and **FMVSS 108** [3, 4, 5].

Construction Vehicle

A construction vehicle, or a construction machine or equipment, is a specialized commercial vehicle designed and built for various construction tasks and operations. These vehicles are essential in the construction industry, where they are employed to excavate, transport, lift, compact, and perform other critical functions required to develop buildings, infrastructure, and civil engineering projects.

Conventional Cab

A conventional cab, often referred to as a conventional or conventional tractor, is a typical style of commercial vehicle cab design used primarily in heavy-duty trucks, such as semi-trucks and tractor-trailers. It is characterized by its recognizable configuration, with a separate front-engine compartment and a passenger compartment behind it. This design allows for better engine accessibility, driver comfort, and aerodynamic performance.

Coolant

Coolant, often referred to as engine coolant or antifreeze, is a specially formulated liquid used in commercial vehicles to regulate the engine's temperature and prevent it from overheating. It serves as a heat transfer medium, ensuring the engine operates within an optimal temperature range while also protecting against freezing in cold weather.

Cooling System

A coolant system, also known as a cooling system or engine cooling system, is an integral component of commercial vehicles that manages the engine's temperature to prevent overheating. It consists of various components and fluids designed to regulate and dissipate excess heat generated during engine operation, ensuring optimal performance and engine longevity.

Cotton Picker

A cotton picker is a specialized agricultural machine designed for the mechanical harvesting of cotton from cotton plants (Figure C.4). These machines are essential in modern cotton farming, as they significantly increase the efficiency of cotton harvesting, reduce labor costs, and improve overall crop yield.

FIGURE C.4 An example of a cotton picker harvesting.

casa.da.photo/Shutterstock.com.

Cotton pickers have rotating spindles or picking units that strip the cotton bolls (seed capsules) from the cotton plants. The cotton bolls are collected into a conveyor system or basket, where they are initially cleaned and separated from leaves, stems, and other plant debris. Cotton pickers have built-in cleaning mechanisms that remove remaining plant material, such as leaves and twigs, from the harvested cotton before it is deposited into the onboard storage modules. The harvested cotton is deposited into large, modular storage containers, often called modules, which are then transported to a processing facility for ginning.

CPM (Cost Per Mile)

CPM is a financial metric used in the commercial vehicle industry to quantify the expenses associated with operating a vehicle over a specific distance traveled. It is a fundamental measure that helps businesses and fleet managers assess their operations' economic efficiency and profitability by breaking down costs per mile.

Calculation of CPM

$$\text{CPM} = \frac{\text{Total costs}}{\text{Total miles}}$$

Total costs

$$\text{Total cost} = \Sigma\left(\cos t_1 + \cos t_2 + \cos t_n\right)$$

The expense associated with purchasing fuel is a significant component of a vehicle's operating costs. Costs are used for routine maintenance, repairs, and occasional replacements of vehicle components. The reduction in the vehicle's value over time is due to wear and tear or depreciation. Premiums are paid to insure the vehicle against various risks and liabilities. Fees are paid for obtaining necessary permits, licenses, and vehicle registration. The cost of tires and expenses are related to purchasing, mounting, and replacing tires, which is significant for over-the-road vehicles. If the vehicle was financed, the interest on the loan was used to purchase it. Compensation is paid to drivers for their services. Fees are paid for using toll roads and various taxes related to the vehicle's operation.

Crane

A crane is a versatile and specialized industrial machine for lifting, lowering, and transporting heavy objects or materials horizontally. Cranes come in various types, configurations, and sizes, each designed for specific construction, manufacturing, shipping, and material handling applications.

For more information, see **J1028** and **J999** [6, 7].

Crane Truck

A crane truck, also known as a truck-mounted crane or mobile crane, is a specialized commercial vehicle that combines the functionality of a truck with an integrated crane apparatus. These versatile vehicles are used in various industries for lifting and moving heavy loads in diverse settings, including construction, transportation, logistics, and material handling.

The defining feature of a crane truck is its built-in crane, which is mounted on the truck's chassis. This crane can lift, lower, and maneuver heavy objects. Depending on the intended use and industry requirements, crane trucks come in different chassis configurations, such as flatbeds, service bodies, or specialized truck types. The crane typically features a telescopic boom extending to various lengths, allowing it to reach different heights and distances.

Crane trucks have specified lifting capacities, which vary depending on the size and design of the crane. Relying on the model, these capacities range from a few tons to several tons or more. Most crane truck cranes are operated hydraulically, offering precise control over the lifting, lowering, and positioning of loads. To maintain stability while lifting heavy loads, crane trucks often have outriggers or stabilizers that can be extended to create a broader support base (Figure C.5).

FIGURE C.5 An example of vehicle layout with outriggers [6].

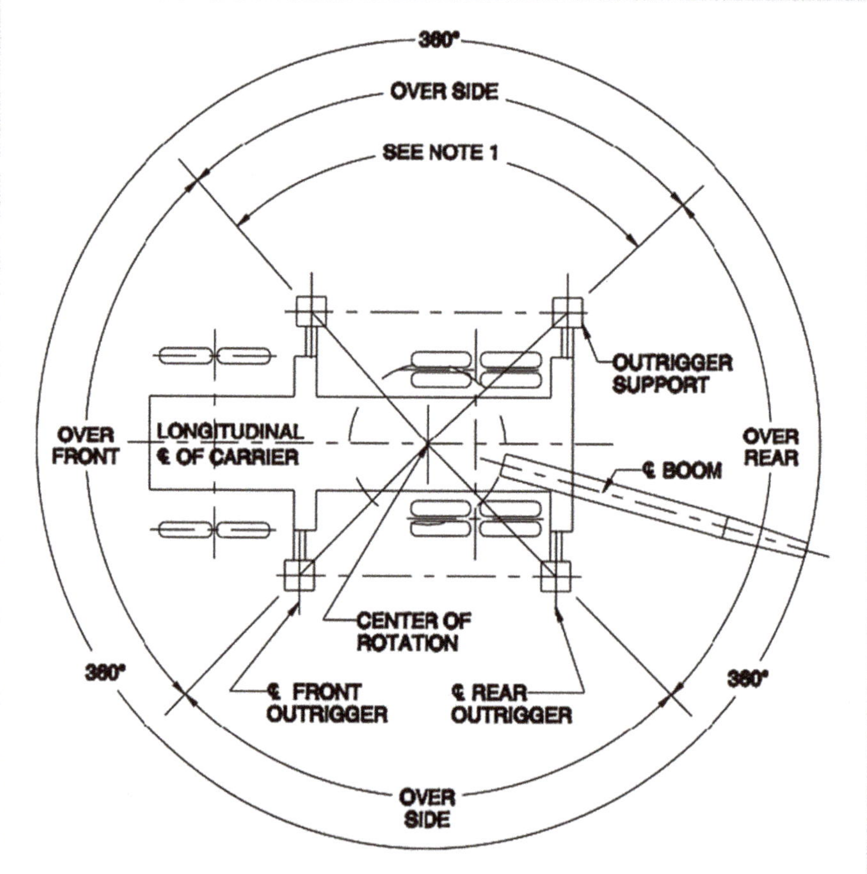

Operators control the crane from a dedicated cabin on the crane or a remote control station, allowing for safe and precise maneuvering. Crane trucks can have various attachments, such as hooks, winches, and specialized lifting tools, to accommodate different load types and configurations. Crane trucks are used in numerous applications, including construction (e.g., lifting construction materials and equipment), logistics (e.g., loading and unloading cargo containers), tree services (e.g., removing tree branches or stumps), and utility maintenance (e.g., repairing power lines).

For more information, see **J1028** [6].

Crawler Loader

A crawler loader is a specialized commercial vehicle in construction, excavation, and material handling applications. It combines the features of a tracked crawler tractor and a front-end loader, offering versatility, stability, and mobility in challenging terrains and worksites.

Crawler loaders have a robust, tracked undercarriage, similar to bulldozers or excavators. This design provides excellent traction, stability, and weight distribution, making them suitable for soft or uneven terrains. At the front of the crawler loader is a loader attachment, which consists of a bucket or scoop that can be raised, lowered, tilted, and articulated. This loader attachment allows for the lifting and carrying of various materials, such as soil, gravel, rocks, and debris. Crawler loaders are versatile machines that multiple transports use to perform various tasks, including excavation, backfilling, grading, digging, loading, and transporting materials.

Crawler loaders use hydraulic systems to control the movements of the loader attachment, providing precise and responsive operation. Many crawler loaders can have various attachments to adapt to specific tasks, such as forks, rakes, blades, and grapples. The wide, tracked undercarriage of crawler loaders helps distribute weight evenly, reducing ground pressure and minimizing damage to sensitive surfaces, making them suitable for wetlands, forests, and other environmentally sensitive areas. Crawler loaders have excellent maneuverability due to their tracked undercarriage, making them ideal for tight spaces and confined worksites. Crawler loaders come in various sizes and models, with lifting capacities ranging from a few tons to several tons, depending on the specific machine.

Cruise Control

Cruise control is a convenient and common automotive feature designed to automatically maintain a constant speed for a vehicle, typically on highways or open roads, without requiring the driver to apply the throttle pedal continuously. This technology enhances driving comfort, reduces driver fatigue, and can contribute to fuel efficiency.

CTIS (Central Tire Inflation System)

A CTIS is a specialized technology installed in particular commercial and off-road vehicles to allow the driver to simultaneously adjust the tire pressure of all the vehicle's tires from within the cabin. This system provides several advantages, including improved traction, increased off-road capability, fuel efficiency, and tire life.

For more information, see **J2848** [8].

Cubic Capacity

Cubic capacity, often referred to as engine displacement or simply displacement, is a critical specification used to describe the size and capacity of an ICE, particularly in commercial vehicles. It indicates the total volume of all the engine's cylinders, typically measured in liters (L) or cubic centimeters (cc).

Cubic capacity represents the total volume of all the engine's cylinders, where each cylinder is essentially a hollow tube with a piston inside. Cubic capacity is usually expressed in L (for larger engines) or cc (for smaller engines). One liter is equivalent to 1000 cc. Cubic capacity is calculated by multiplying the cross-sectional area of a single cylinder's bore (the diameter of the cylinder) by the stroke length (the distance the piston travels within the cylinder) and then multiplying this by the number of cylinders in the engine. The formula is typically expressed as:

$$\text{Cubic feet} = \text{Length}\left(\text{ft}\right)\times\text{Width}\left(\text{ft}\right)\times\text{Height}\left(\text{ft}\right)$$

CVAR (Commercial Vehicle Autonomous Routing)

CVAR box is an advanced technology solution integrated into commercial vehicles to facilitate autonomous routing and navigation. This innovative system combines navigation, real-time data processing, and connectivity to enhance commercial vehicle operations' efficiency, safety, and autonomy.

The CVAR box has advanced navigation and mapping software that utilizes GPS, satellite data, and mapping databases to determine optimal routes for commercial vehicles. The CVAR box continuously processes real-time data, including traffic conditions, road closures, weather forecasts, and real-time vehicle data, to adjust routes and make informed decisions dynamically. The CVAR box connects external data sources, such as traffic management systems and fleet management platforms, to receive and transmit relevant information. Using sophisticated algorithms, the CVAR box calculates routes considering vehicle type, weight, dimensions, road restrictions, and delivery schedules.

Drivers interact with the CVAR box through a user-friendly interface that displays route information and real-time updates and suggested alternate routes.

The CVAR box enables vehicle communication and fleet management systems, allowing real-time updates on vehicle status, location, and route deviations.

The CVAR box streamlines commercial vehicle operations by providing the most efficient and time-saving routes, reducing travel time and operational costs. By analyzing real-time traffic data, the CVAR box helps drivers avoid congested areas, decreasing delays and improving overall route efficiency. The CVAR box considers road conditions, weather, and traffic incidents to create safer routes, minimize the risk of accidents, and ensure driver well-being.

REFERENCES

1. Society of Automotive Engineers, "J2185 Life Test for Heavy-Duty Storage Batteries (Lead Acid Type Only)," SAE Publishing, Warrendale, 2018.
2. Bullough, J., "Intelligent Warning Lights and Driving Safety," SAE Technical Paper 2015-01-1700 (2015), doi:https://doi.org/10.4271/2015-01-1700.
3. Society of Automotive Engineers, "J1967 Retroreflective Materials for Vehicle Conspicuity," SAE Publishing, Warrendale, 2006.
4. Society of Automotive Engineers, "J759 Lighting Identification Code," SAE Publishing, Warrendale, 2017.
5. National Highway Traffic Safety Administration, "Code of Federal Regulations: Federal Motor Vehicle Safety Standards," January 9, 2024, accessed January 12, 2024, https://www.ecfr.gov/current/title-49/subtitle-B/chapter-V/part-571/subpart-B/section-571.108.
6. Society of Automotive Engineers, "J1028 Mobile Crane Working Area Definitions," SAE Publishing, Warrendale, 2022.
7. Society of Automotive Engineers, "J999 Crane Boom Hoist Disengaging Device," SAE Publishing, Warrendale, 2022.
8. Society of Automotive Engineers, "J2848 Tire Pressure Systems - Management (CTIS) Type for Medium- and Heavy-Duty Highway Vehicles," SAE Publishing, Warrendale, 2021.

D

Data Integration

Data integration involves combining GPS data with other sources of information, such as weather data and crop history, to make informed planting decisions.

Data Logging

Data logging records GPS and planting-related data for analysis and decision-making in future planting seasons.

Day Cab

A day cab, short for day cabin or day cab tractor, is a type of commercial vehicle configuration commonly found in the trucking industry. It is designed explicitly for short-haul or regional transportation, where the driver is not expected to sleep in the vehicle overnight. Day cabs are typically characterized by compact size, lacking a sleeper compartment standard in long-haul trucks.

DCT (Dual-Clutch Transmission)

A DCT, a twin-clutch transmission or a direct-shift gearbox (DSG), is an advanced transmission system used in some commercial vehicles. DCTs offer rapid and efficient gear shifts by utilizing two clutch assemblies, each responsible for engaging and disengaging specific gears. This design allows for seamless and near-instantaneous gear changes, enhancing vehicle performance and fuel efficiency.

The defining feature of a DCT is the presence of two clutch assemblies. One clutch is responsible for odd-numbered gears (1st, 3rd, 5th, etc.), and the other handles even-numbered gears (2nd, 4th, 6th, etc.).

DCTs can pre-select the next gear, while the current gear is engaged. This enables almost instantaneous gear changes by engaging one clutch while disengaging the other, resulting in smoother and quicker shifts. DCTs often offer both manual and automatic driving modes. In manual mode, the driver can select gears manually using paddle shifters or a gear lever, while automatic mode allows the transmission to shift gears automatically.

DCTs are designed to improve fuel efficiency by reducing energy losses associated with traditional torque converters in automatic transmissions. The rapid and seamless gear changes in DCTs contribute to improved acceleration and overall driving performance.

DEF (Diesel Exhaust Fluid)

DEF is a specialized aqueous solution used in modern commercial vehicles equipped with SCR technology to reduce harmful emissions from diesel engines. DEF is crucial in reducing NOx emissions, a significant contributor to air pollution.

DEF is composed of purified water (H_2O) and urea, a chemical compound that helps reduce NOx emissions when injected into the exhaust stream of a diesel engine. DEF is an integral part of the SCR system, designed to considerably reduce NOx emissions by converting them into harmless nitrogen gas and water vapor. This process occurs within the SCR catalyst in the exhaust system.

DEF is typically stored in a dedicated tank on the commercial vehicle. It should be handled carefully, kept in clean and sealed containers, and protected from extreme temperatures to maintain effectiveness. DEF is injected into the exhaust system upstream of the SCR catalyst using a precise dosing system. This ensures the correct ratio of DEF to exhaust gases for efficient NOx reduction.

The rate of DEF consumption varies depending on the engine's NOx emissions, which in turn relies on factors such as engine load, operating conditions, and vehicle usage. Commercial vehicles have DEF-level sensors to monitor and indicate when a refill is needed. Many regions and countries have implemented strict emission regulations for commercial vehicles. DEF and SCR technology are shared compliance methods, helping vehicles meet emission standards like Euro 6 and Environmental Protection Agency (EPA) 2010. The reduction of NOx emissions achieved through DEF and SCR technology has a positive environmental impact, improving air quality and reducing harm to ecosystems.

DEF must meet specific quality standards, such as the International Organization for Standardization (ISO) 22241, to ensure its effectiveness and compatibility with SCR systems. Using subpar DEF can lead to engine performance issues and emission non-compliance. DEF is readily available at fueling stations, truck stops, and automotive supply stores, making it convenient for commercial vehicle operators to refill their DEF tanks. Regular maintenance includes checking DEF levels, ensuring proper storage and handling, and maintaining the SCR system to assure its continued effectiveness.

Delivery Truck

A delivery truck, often called a delivery van or delivery vehicle, is a specialized commercial vehicle designed primarily for transporting and delivering goods and cargo to various destinations. These trucks are crucial for the logistics and distribution industry, serving as the backbone for the delivery of products to businesses, retailers, homes, and other locations.

Delivery trucks are characterized by their enclosed cargo compartments, which protect goods from weather conditions, theft, and damage during transport. Depending on the specific model and intended use, the cargo area may vary in size and configuration. Delivery trucks come in various sizes, from small vans to larger box trucks. The size of the vehicle and its cargo capacity are selected based on the volume and type of cargo being transported. The payload capacity of a delivery truck represents the maximum weight of cargo and goods it can carry safely. This capacity is determined by the vehicle's design, axles, suspension, and legal weight limits.

Delivery trucks typically have rear cargo doors for loading and unloading. Depending on the model and manufacturer, these doors may be swing-out, roll-up, or sliding. Some trucks have additional side doors for easier access to the cargo. Delivery trucks can have different configurations, including panel vans, step vans, box trucks, and refrigerated vans, each designed for specific cargo and delivery requirements.

Many delivery trucks are customizable to meet specific delivery needs. This may include interior shelving, refrigeration units, or specialized compartments for delicate or perishable goods. Some delivery trucks are equipped with loading ramps or liftgates to facilitate the loading and unloading of heavy or oversized items. The driver's cabin of a delivery truck is furnished with ergonomic seating, controls, and visibility features to ensure operator comfort and safety during long delivery routes.

Modern delivery trucks are designed with fuel efficiency in mind to reduce operating costs and environmental impacts. This includes using aerodynamic designs and fuel-efficient engines. Delivery trucks are standard in urban areas, where they navigate city streets and neighborhoods to make deliveries to businesses and residential addresses.

Courier and parcel delivery companies frequently use them to transport packages, mail, and other small- to medium-sized shipments. Some delivery trucks have refrigeration or temperature-controlled units to transport perishable goods, such as groceries, pharmaceuticals, and flowers, at specific temperatures.

Delivery Vehicle

See **Delivery Truck**.

Desiccant

A desiccant and wet tank system in commercial vehicles refers to a combination of components used to manage moisture and contaminants in the air brake system. The system typically consists of a desiccant air dryer and a wet tank, and it plays a crucial role in maintaining the integrity and reliability of the vehicle's braking system.

For more information, see **J2383** [1].

DF1 (Diesel Fuel #1)

DF1 or winter diesel is a specific diesel fuel used primarily in regions with frigid temperatures. It is a lighter and more refined diesel fuel than Diesel Fuel #2 (DF2), the standard diesel fuel used in most regions. DF1 is formulated to perform effectively in very low temperatures, preventing issues such as gelling and waxing when regular diesel fuel is exposed to extreme cold.

DF1 is a lighter distillate fuel, refined to remove heavier components such as waxes and impurities. It has a lower boiling point and is less dense than DF2. The primary purpose of DF1 is to provide reliable engine performance in icy weather conditions. It has a lower cloud point (the temperature at which wax crystals begin to form) and a lower pour point (the temperature at which the fuel becomes semisolid) than DF2. In frigid temperatures, regular diesel fuel (DF2) can form gel-like substances due to the solidification of waxes in the fuel. DF1 is less prone to this issue, making it suitable for regions where temperatures frequently drop below freezing.

Diesel engines can be challenging to start in freezing weather due to the thickening of the fuel. DF1 flows more easily at low temperatures, ensuring the engine starts smoothly even in sub-zero conditions. While DF1 offers superior cold weather performance, it has a slightly lower energy content than DF2. This means that it may result in slightly reduced fuel efficiency and power output in warmer conditions.

DF1 is primarily used in regions with harsh winter climates, such as northern parts of North America and Europe, where cold weather can cause regular diesel fuel to become problematic. In some regions, a seasonal practice involves blending DF1 with DF2 to create a winter-grade diesel fuel. This blend improves cold weather performance without completely switching to DF1. Most diesel engines designed to run on DF2 can also operate on DF1. However, it is essential to consult the manufacturer's recommendations for specific fuel requirements, especially if switching between fuel types (Figure D.1).

FIGURE D.1 Fuel oil composition per J313 [2].

	No. 1-D S15	No. 1-D S500	No. 1-D S5000	No. 2-D S15	No. 2-D S500	No. 2-D S5000	No. 4-D
Flash Point, °C, min	38	38	38	52	52	52	55
Water and Sediment, volume %, max	0.05	0.05	0.05	0.05	0.05	0.05	0.50
Carbon Residue, 10% Residuum, % mass, max	0.15	0.15	0.15	0.35	0.35	0.35	---
Ash, % mass, max	0.01	0.01	0.01	0.01	0.01	0.01	0.10
Distillation Temperature °C, 90% vol. recovered, min	---	---	---	282	282	282	---
Distillation Temperature °C, 90% vol. recovered, max	288	288	288	338	338	338	---
Kinematic Viscosity, mm²/s [4], at 40 °C, min	1.3	1.3	1.3	1.9	1.9	1.9	5.5
Kinematic Viscosity, mm²/s [4], at 40 °C, max	2.4	2.4	2.4	4.1	4.1	4.1	24.0
Sulfur [5], % mass, max	0.0015	0.05	0.50	0.0015	0.05	0.50	2.00
Copper Strip Corrosion, rating, max	No. 3	No. 3	No. 3	No. 3	No. 3	No. 3	---
Cetane Number [6], min	40 [9]	40 [9]	40 [9]	40 [9]	40 [9]	40 [9]	30 [8]
Cloud Point, °C, or LTFT/CFPP, °C, max	[7]	[7]	[7]	[7]	[7]	[7]	
Lubricity, HFRR @ 60 °C, micron, max	520	520	520	520	520	520	---
Conductivity, pS/m, min	25 [12]	25 [12]	25 [12]	25 [12]	25 [12]	25 [12]	---

(1) To meet special operating conditions, modifications of individual limiting requirements may be agreed upon between purchaser, seller, and manufacturer.

(2) The values in SI units are to be regarded as the standard. The values of U.S. customary units are for information only.

(3) Nothing in this specification shall preclude observance of federal, state, or local regulations which may be more restrictive.

(4) Millimeter squared per second Note: 1 centiStokes (cSt) = 1 mm²/s).

(5) In countries outside the USA, other sulfur limits may apply.

(6) Where cetane number by Method D613 is not available, Method D4737 may be used as an approximation. Where there is disagreement, Method D613 shall be the referee method.

DF2 (Diesel Fuel #2)

DF2 is the most common and widely used diesel fuel in the transportation and commercial vehicle industry. It is a versatile and efficient fuel that powers various diesel engines, from trucks and buses to construction equipment and generators. DF2 is characterized by its energy density, relatively high cetane rating, and widespread availability.

DF2 is a distillate fuel derived from crude oil through refining. It contains hydrocarbon (HC) molecules that are heavier and denser than gasoline ones. DF2 has a high energy density, meaning it contains significant energy per unit volume. This high energy content gives diesel engines excellent fuel efficiency and power output. DF2 typically has a higher cetane rating than other diesel fuels. Cetane is a measure of a diesel fuel's ignition quality. A higher cetane rating indicates better ignition characteristics and smoother engine operation.

For more information, see **J313** [2].

Diesel Fuel Heater

A diesel fuel heater is a specialized device designed to heat diesel fuel before it enters a vehicle's engine. It is commonly used in commercial vehicles, especially in regions with cold climates, to ensure the smooth and reliable operation of diesel engines, even in extremely low temperatures.

In frigid temperatures, diesel fuel can thicken and form gel-like substances due to the solidification of waxes present in the fuel. A diesel fuel heater prevents this gelling by heating the fuel before it reaches the engine. Diesel engines can be challenging to start in freezing weather due to the thickening of the fuel. A fuel heater aids in the cold-start process by providing preheated fuel to the engine, improving combustion, and ensuring the engine starts smoothly. A typical diesel fuel heater consists of several components, including a heating element (often electrical or using combustion), a fuel pump to circulate the fuel through the heater, and a control system to regulate the heating process.

Diesel fuel heaters are typically integrated into the vehicle's fuel system, either as an aftermarket addition or as part of the original equipment in some commercial vehicles. Many diesel fuel heaters have temperature control settings that allow the operator to adjust the level of fuel heating based on the prevailing weather conditions. This ensures that the fuel remains within the appropriate temperature range for the engine.

Diesel fuel heaters are designed to be energy-efficient, heating only the fuel needed to maintain the desired temperature. This helps minimize energy consumption and improve fuel efficiency. Safety features are often incorporated into diesel fuel heaters to prevent overheating, overpressure, and other potential issues. These features may include temperature sensors and automatic shut-off mechanisms. Diesel fuel heaters are compatible with various types of diesel fuel, including standard DF2 and winter-grade diesel fuels. They are designed to accommodate different fuel viscosities.

Diesel Knock

Diesel knock, also known as diesel engine knock, is an audible sound characterized by a sharp, metallic tapping or knocking noise that can occur in diesel engines used in commercial vehicles. It typically happens during combustion and is associated with fuel ignition within engine cylinders.

Diesel Particle Filter

See **Diesel Particulate Filter.**

Diesel Plunger Pump

A diesel plunger pump is a type of high-pressure fuel injection pump used in diesel engines of commercial vehicles. This pump is critical in delivering precisely metered and pressurized fuel to the engine's cylinders for combustion, enabling efficient and powerful performance.

1. Fuel Injection: The primary function of a diesel plunger pump is to inject diesel fuel into the engine's cylinders at high pressure and with precise timing. This injection process is essential for the combustion of fuel within the engine.
2. High Pressure: Diesel plunger pumps are designed to generate high-pressure fuel delivery, ensuring that the fuel atomizes effectively and combusts efficiently in the engine cylinders.
3. Plunger Mechanism: The pump features a plunger and barrel assembly, where the plunger moves within a cylindrical barrel to compress and deliver fuel in a controlled manner.
4. Timing and Metering: The pump's timing and metering are critical for delivering the right amount of fuel at the exact moment needed during each engine cycle, contributing to optimal performance and emission control.
5. Governor Control: Some diesel plunger pumps are equipped with a governor, which controls the fuel delivery rate based on engine speed and load, allowing for smooth operation and fuel efficiency.

Differential

A differential is a critical component in the drivetrain of commercial vehicles, serving as a mechanical device that enables the distribution of power between the wheels on the same axle while allowing them to rotate at different speeds. It is fundamental in ensuring smooth and effective vehicle operation, particularly during turns (Figure D.2).

FIGURE D.2 Differential removal from a modern automatic transmission.

PBabic/Shutterstock.com.

1. Power Distribution: The primary function of a differential is to distribute engine power from the driveshaft to the wheels of the same axle, allowing each wheel to receive the appropriate amount of power.
2. Rotation at Different Speeds: When a vehicle makes a turn, the outer wheel needs to cover a longer distance than the inner wheel. The differential allows the wheels to rotate at different speeds during this maneuver, preventing tire scrubbing and wear.
3. Planetary Gears: Most differentials utilize a set of planetary gears to achieve power distribution and speed differentiation. These gears ensure that the wheels receive the appropriate amount of torque while allowing for relative motion.
4. Locked and Open Differentials: Commercial vehicles can be equipped with either open or locked differentials. Open differentials allow for speed differentiation between wheels, while locked differentials force both wheels to turn at the same speed, improving traction in challenging conditions.

5. Limited-Slip Differential: Some commercial vehicles feature limited-slip differentials, which balance open and locked differentials, enabling varying degrees of wheel speed differentiation while maintaining some level of traction between wheels.

Direct Drive

Direct drive in commercial vehicles refer to a type of powertrain configuration in which the engine's power is transmitted directly to the vehicle's drivetrain without using a secondary transmission or gearbox. In this setup, the engine's crankshaft is mechanically connected to the vehicle's wheels, providing a straightforward and efficient power transfer.

Direct Injection

Direct injection, often referred to as direct fuel injection, is an advanced technology used in the fuel delivery systems of many modern commercial vehicles, particularly diesel and gasoline engines. It is designed to improve fuel efficiency, reduce emissions, and enhance engine performance by precisely injecting fuel directly into the engine's combustion chamber.

Direct injection systems enable precise and controlled fuel delivery directly into the engine's combustion chamber, bypassing the intake manifold. Fuel is delivered at high pressure through fuel injectors, ensuring it atomizes effectively and mixes thoroughly with air for efficient combustion. Direct injection systems often allow for multiple fuel injections during a single engine cycle, optimizing combustion and emission control. Direct injection is used in both diesel and gasoline engines, although the technology and components may vary slightly between the two. Direct injection is controlled by the vehicle's ECU, which adjusts the timing and duration of fuel injection based on real-time data, such as engine load, speed, and emission requirements.

Disc Brake

A disc brake is a type of braking system commonly used in commercial vehicles to slow down or stop the vehicle. It consists of a flat, rotating disc (the brake rotor) and a set of friction-inducing brake pads that clamp onto the rotor when the brakes are applied. Disc brakes offer effective and consistent stopping power, making them popular for various commercial vehicle applications.

For more information, see **J2932** [3].

Disc Harrows

Disc harrows are a type of agricultural implement commonly used in farming and soil preparation (Figure D.3). They are designed to break up and cultivate soil by cutting and turning it using a series of circular metal discs mounted on a frame. Disc harrows are vital tools for seedbed preparation, weed control, and improving soil structure in agriculture.

FIGURE D.3 Disc harrows attach to the tractor.

Michael Dechev/Shutterstock.com.

Disc harrows consist of circular metal discs, often made of hardened steel, that are mounted on axles or gangs. These discs have sharp edges and may vary in size and spacing depending on the specific harrow design and intended use. Disc harrows are attached to a frame with a hitch mechanism, which can be connected to a tractor or other suitable towing vehicle. The frame provides stability and adjusts the harrow's working depth and angle. As the harrow is pulled through the field, the rotating discs cut into the soil, breaking it into smaller clumps and turning it over. This process effectively incorporates crop residues, organic matter, and amendments into the soil. Disc harrows can be adjusted to control the depth at which they work the soil. This flexibility allows farmers to prepare the soil at the desired depth for planting or seeding.

Disc harrows are commonly used to prepare seedbeds before planting crops. They create a fine, loose soil structure conducive to seed germination and root development. By cutting and burying weeds and weed seeds, disc harrows help control weed populations and prevent crop competition. The action of disc harrows helps aerate the soil by breaking up compacted layers and improving H_2O infiltration and root penetration. Disc harrows effectively incorporate crop residues, such as stalks and leftover plant material from previous crops, into the soil. This aids in decomposition and nutrient recycling.

Disc harrows come in various types and configurations, including tandem disc harrows (with two gangs of discs), offset disc harrows (for more aggressive tillage), and vertical tillage disc harrows (designed for minimal soil disturbance). Regular maintenance of disc harrows is essential to ensure proper operation. This includes lubricating bearings, sharpening or replacing worn discs, and inspecting the frame and hitch for wear and damage. Disc harrows are typically attached to tractors with compatible power and hitching systems. The size and type of harrow chosen should match the tractor's horsepower and the specific soil preparation requirements.

Disc Mowers

Disc mowers are agricultural machinery used for efficiently cutting and conditioning grass or hay in commercial farming and forage production. They are designed to replace traditional sickle-bar mowers with a more modern and effective cutting mechanism. Disc mowers consist of rotating circular blades (discs) that swiftly and cleanly cut standing crops, making them a popular choice for haymaking and forage harvesting.

Disc mowers are equipped with multiple round cutting discs, typically two or more, mounted on a horizontal cutter bar. These discs rotate at high speeds to cut the standing crop cleanly. The cutting discs are fitted with sharp, replaceable blades or knives. The blades are designed to create a precise and uniform cut, reducing crop damage and promoting faster drying. Disc mowers can be adjusted to control the cutting height, allowing farmers to customize the crop length and quality. This flexibility is essential for different types of forage and harvesting conditions. Disc mowers have a wide cutting swath, which means they can cover a significant width in a single pass. This results in increased productivity and reduced time and labor required for mowing. The high rotation speed of the cutting discs ensures rapid and efficient cutting of grass or crops. This speed contributes to quick harvesting and minimizes the risk of crop lodging (plants falling over).

Disc mowers deliver a consistent and uniform cut across the entire width of the swath. This uniformity is crucial for producing high-quality forage and hay with even drying and nutritional content. Some disc mowers have a conditioning system, such as rubber rollers or flail tines, that crimps or crushes the cut crop as it passes through. Conditioning accelerates drying and improves forage quality. Disc mowers are typically attached to tractors or other towing vehicles using a three-point hitch or other suitable attachment methods. They are compatible with a range of tractor sizes and power ratings.

Disc mowers are commonly used for harvesting various forage crops, including grass, alfalfa, clover, and other legumes. They are essential tools in hay production and forage harvesting operations. Many disc mowers can be raised to a transport position for easy and safe transportation between fields and storage areas. This reduces the risk of damage during transit.

Dispatcher

A dispatcher is an individual or a system responsible for coordinating and managing the movement, scheduling, and deployment of a fleet of commercial vehicles, such as trucks, vans, buses, or delivery vehicles. The primary role of a dispatcher is to optimize the efficiency of vehicle operations, ensure timely deliveries, and maintain effective communication between drivers, vehicles, and customers.

Dispatchers create optimal routes and schedules for vehicles based on delivery locations, traffic conditions, vehicle capacity, and time constraints. This helps in maximizing the utilization of resources and minimizing travel time. Dispatchers serve as a central point of communication between drivers, customers, and other stakeholders. They relay important information, instructions, and updates to drivers regarding route changes, delays, or special requests.

Using GPS tracking and fleet management software, dispatchers monitor each vehicle's real-time location and status. This allows them to address unexpected issues, adjust routes, and provide accurate delivery estimates to customers.

In case of emergencies, accidents, breakdowns, or other unforeseen situations, dispatchers coordinate the appropriate response, such as sending assistance, redirecting vehicles, or informing authorities.

Dispatchers analyze data and use software tools to optimize routes and operations. This includes fuel efficiency, load balancing, and reducing idle time. Dispatchers interact with customers to provide updates on delivery status, address inquiries, and handle any issues or concerns that may arise during delivery. Dispatchers maintain records of routes taken, delivery times, vehicle maintenance, and other relevant data. This documentation helps evaluate performance, identify trends, and make informed decisions. Dispatchers allocate vehicles and drivers to specific jobs or routes based on availability, skills, and requirements.

Displacement

Displacement, often called engine displacement or engine capacity, is a crucial specification in commercial vehicles that measures the total volume of air–fuel mixture an engine can draw into its cylinders during a complete engine cycle. It is typically expressed in liters (L) or cubic centimeters (cc) and significantly determines the engine's power output and performance.

DOC (Diesel Oxidation Catalyst)

A DOC is an emission control device installed in the exhaust system of commercial vehicles with diesel engines. It is designed to reduce harmful emissions by promoting the oxidation of exhaust gases, mainly carbon monoxide (CO) and HCs, into less toxic compounds, such as carbon dioxide (CO_2) and H_2O.

DOF (Degrees of Freedom)

DOF in commercial vehicles refers to the independent parameters or variables that define the spatial movement and flexibility of a vehicle's components, such as wheels, axles, and suspension systems. It is important in designing, analyzing, and controlling commercial vehicle dynamics and maneuverability.

Dog Clutch

A dog clutch, also known as a dog gear or sliding gear, is a mechanical component used in commercial vehicles and other machinery with manual transmissions. It plays a pivotal role in engaging and disengaging gears, allowing the driver to select different speeds or directions for the vehicle.

1. Gear Engagement: The dog clutch is responsible for connecting and disconnecting gears within a manual transmission. It enables the driver to choose different gear ratios for driving, such as shifting between forward and reverse gears or varying speeds.
2. Simple Design: Dog clutches have a relatively simple and robust design, consisting of two interlocking gears with matching teeth or "dogs" that can be slid into engagement when needed.
3. Manual Operation: In commercial vehicles with manual transmissions, the engagement and disengagement of the dog clutch are manually controlled by the driver using the gear shifter and the clutch pedal.
4. Synchronization: In modern transmissions, synchronization mechanisms are often incorporated into the dog clutch to facilitate smoother gear changes and prevent grinding when shifting.

Doghouse

A doghouse refers to a compact housing or enclosure typically located in the cab or interior of a vehicle, often used to house equipment, controls, or components related to the vehicle's operation or specific functions.

A doghouse is a small enclosure designed to fit within the limited space of a vehicle's cab or interior. It provides easy access to essential controls, switches, or equipment without taking up excessive space. The enclosure protects sensitive equipment or controls from damage, dirt, moisture, and other environmental factors. The doghouse may house various components, such as electrical systems, hydraulic controls, instrumentation, or communication equipment.

Dolly Truck

A dolly truck, often referred to simply as a dolly, is a specialized type of trailer or equipment used in commercial transportation to enhance larger vehicles' capacity, flexibility, and maneuverability, particularly semi-trucks and tractor–trailers. Dolly trucks are typically designed to support the front or rear portion of a semi-trailer or full trailer, allowing the combination to carry more cargo or providing increased maneuverability for specific applications.

A dolly truck consists of one or more axles equipped with wheels that support the front or rear section of a trailer. These axles may be steerable or non-steerable, depending on the design. Dolly trucks are equipped with a coupling mechanism that attaches to the towing vehicle (typically a tractor or another truck). This mechanism secures the dolly to the towing unit and allows it to support the additional trailer. Dolly trucks are designed to distribute the weight of the additional trailer's cargo, enhancing the overall payload capacity of the combination vehicle. In cases where additional maneuverability is required, dolly trucks with steerable axles can assist the vehicle in making tight turns or navigating challenging road conditions. Dolly trucks come in various configurations, such as single-axle dollies, tandem-axle dollies, and even triple-axle dollies, depending on the specific application and weight requirements.

Double-Decker Bus

Double-decker buses refer to large motor vehicles designed with two levels of passenger seating (Figure D.4). These buses are characterized by their unique design, which allows them to accommodate more passengers efficiently using vertical space.

FIGURE D.4 The iconic double-decker bus has many uses including tourism.

James Steidl/Shutterstock.com.

Double-decker buses have two distinct levels of passenger seating, with an upper deck positioned above the lower deck. Double-decker buses can accommodate more passengers than single-level buses due to their dual-level design. Passengers access the upper deck using stairs inside the bus or, in some cases, lifts for enhanced accessibility.

Depending on the model and purpose, double-decker buses may be equipped with air conditioning, seating configurations, and entertainment systems. Double-decker buses often feature a tall and distinct exterior design with large windows, providing passengers on the upper deck with panoramic views.

Download OTA (Over-the-Air)

OTA, often referred to as OTA updates or remote software updates, is a technology that allows commercial vehicle manufacturers and operators to wirelessly update and upgrade the vehicle's software and firmware, including its control systems and applications. These updates are typically delivered via a wireless network connection, providing a convenient and efficient means of keeping vehicles up-to-date with the latest software improvements.

OTA technology relies on wireless communication, such as cellular networks or dedicated vehicle telematics systems, to deliver software updates to commercial vehicles. OTA allows for the remote installation of software updates, patches, and enhancements for various vehicle systems, including ECUs, infotainment systems, telematics, and safety features. In addition to software, OTA can also update firmware, which includes embedded software that controls hardware components like engine controllers and transmission systems.

OTA systems often incorporate robust security measures to protect the update process from unauthorized access and cyber threats, ensuring the safety and integrity of the vehicle's software. OTA updates provide a convenient means of keeping vehicles up-to-date without requiring physical visits to a service center, reducing downtime and operational disruptions.

DPF (Diesel Particulate Filter)

A DPF is an emission control device integrated into the exhaust system of a diesel engine. Its primary purpose is to reduce the emission of harmful particulate matter (PM), which consists of microscopic soot particles and other solid pollutants produced during the combustion process in diesel engines. DPFs are crucial in modern diesel vehicles to meet stringent emission regulations and improve air quality [4].

A DPF traps and captures PM from the exhaust gases of a diesel engine. It functions like a fine-mesh filter, allowing exhaust gases to pass through while capturing soot particles. DPFs are made of ceramic or metal materials that can withstand the exhaust system's high temperatures and chemical reactions. The filter is designed with a porous structure to facilitate the capture of particles. As soot particles accumulate in the DPF, they can clog the filter and reduce exhaust flow, potentially leading to engine performance issues. To prevent this, DPFs have a regeneration process (Figure D.5).

FIGURE D.5 A table illustrating the soot emission and DPF soot mass change over the RPM range [4].

#	Engine speed (r/min)	Engine torque (N·m)	Exhaust temperature (°C)	NO_2 emission at DPF inlet (ppm)	Soot emission (mg/s)	Test duration (h)	DPF soot mass change (g/L)	Catalytic soot oxidation rate (g/h)
1	1100	110	255	173	1.36	1	0.57	2.75
2	1100	134	291	221	2.19	1	0.02	7.81
3	1100	147	304	294	2.48	1	0.79	11.91
4	1100	153	311	294	2.55	1	1.49	14.79
5	1100	162	322	350	2.06	0.83	1.80	15.49
6	1100	188	383	173	4.95	1	1.21	22.35
7	1900	55	270	17	0.19	0.78	0.14	0.01
8	1900	87	293	60	0.04	1.25	0.25	0.88
9	1900	110	314	88	0.14	1.33	0.97	3.23
10	1900	135	349	113	0.40	1.1	1.08	5.14
11	1900	155	369	96	0.68	1.1	1.44	7.35
12	1900	185	383	115	1.68	0.92	1.90	13.85

There are two primary regeneration methods: passive and active. Passive regeneration occurs during normal driving when exhaust temperatures are high enough to burn off the trapped soot. This process is automatic and requires no driver intervention. Active regeneration is used when passive regeneration is insufficient; the vehicle's ECU initiates active regeneration. This involves injecting extra fuel into the exhaust stream to raise the temperature and burn off the soot. Active regeneration may require periodic maintenance.

A DPF can become saturated with soot, requiring cleaning or replacement over time. Some DPFs have a service interval that needs to be professionally cleaned or replaced. DPFs are highly effective in reducing harmful emissions, including PM, HCs, and CO, as well as some volatile organic compounds (VOCs). This contributes to improved air quality and compliance with emission standards [5].

Many regions and countries have established emission standards that require installation and proper functioning of DPFs in diesel vehicles. Non-compliance can result in fines and vehicle registration restrictions.

Some modern DPF systems are equipped with adaptive controls that optimize regeneration cycles based on driving conditions, load, and exhaust temperatures. This enhances fuel efficiency and reduces emissions. Vehicles with DPFs are equipped with diagnostic systems that monitor the DPF's condition and performance. They may display warning lights or messages to alert the driver or technician of any issues.

For more information, see *Evaluation of Different Methodologies of Soot Mass Estimation for Optimum Regeneration Interval of DPF* [6].

Drag

See **Aerodynamics**.

Drag Coefficient

See **Aerodynamics**.

Drive Axle

A drive axle, also known as a powered axle or drive axle assembly, is a key component in commercial vehicles responsible for transmitting power from the vehicle's engine to the wheels that propel the vehicle forward. It plays a central role in powering the vehicle, and its design can significantly impact the vehicle's performance, traction, and handling.

Drive Cycle

A drive cycle, also known as a vehicle drive cycle, is a standardized testing procedure used in the automotive industry, including commercial vehicles, to evaluate and measure a vehicle's performance, emissions, and fuel efficiency under specific driving conditions. Drive cycles replicate real-world driving scenarios, providing a consistent and repeatable method for assessing a vehicle's performance and compliance with regulatory standards.

Driveline

The driveline, often referred to as the drive train, is a critical component in commercial vehicles responsible for transmitting power from the engine to the wheels that drive the vehicle. It consists of various mechanical components that transfer torque and allow the vehicle to move.

Driver Management System

This system focuses on driver performance monitoring, training, and compliance. It tracks driver behavior, safety metrics, training records, and compliance with regulations.

Driver Performance Monitoring

These systems track driver behavior, including speeding, harsh braking, and rapid acceleration, promoting safer driving habits and reducing accidents.

Driveshaft

A driveshaft, also known as a propeller shaft or prop shaft, is a mechanical component in commercial vehicles that serves as a critical link between the transmission or gearbox and the vehicle's drive axle or axles. Its primary function is to transmit torque and rotational power from the transmission to the wheels, enabling the vehicle to move.

Driveshaft (Propeller Shaft) Angle Measuring

Driveshaft angle measuring, often referred to as propeller shaft angle measurement, is a diagnostic procedure used in commercial vehicles to assess the angles of the driveshaft, especially at the connection points between the transmission, driveshaft, and drive axles. This measurement is crucial for diagnosing and addressing driveline issues and ensuring the efficient and safe operation of the vehicle.

Driveshaft angle measuring is used to determine the angles at which the driveshaft connects to the transmission and the drive axles. This information is critical for assessing the alignment and balance of the driveshaft, which can affect vehicle performance and reduce wear and tear. The procedure measures the angles at the universal joints or U-joints, which are critical for accommodating variations in angles and suspension movement. Proper angles are necessary to prevent U-joint binding or premature wear.

Accurate driveshaft angle measurement helps identify misalignments that may lead to vibrations and noise during vehicle operation. Correcting these angles can mitigate these issues. Driveshaft angle measurement is a valuable tool for maintenance and troubleshooting. If angles are incorrect, it can signal the need for adjustments or component replacement.

Driveshaft (Propeller Shaft) Balancing

Driveshaft balancing, often called propeller shaft balancing, is a maintenance and diagnostic procedure used in commercial vehicles to ensure that the driveshaft rotates smoothly and without excessive vibrations. Balancing is crucial for preventing wear and tear, reducing noise, and improving the overall efficiency and safety of the vehicle's driveline.

Drivetrain

See **Driveline**.

Drop Deck Trailer

A drop deck trailer, also known as a step deck trailer, is a specialized type of semi-trailer commonly used in commercial transportation (Figure D.6). It is characterized by its unique design, featuring two distinct deck levels: an upper deck and a lower deck. This configuration allows for the efficient transportation of cargo that may be oversized or challenging to load onto standard flatbed trailers.

FIGURE D.6 The drop deck trailer is used to carry large loads effectively.

Hmong Windows/Shutterstock.com.

The most prominent feature of a drop deck trailer is its two-level design. The upper deck is typically shorter than the lower deck, and the two decks are connected by a set of steps or ramps, which give the trailer its name.

The lower deck of the drop deck trailer is closer to the ground, making it easier to load and unload cargo that may be too tall to fit on a standard flatbed trailer or require a lower clearance for safe transportation.

Drop deck trailers are highly versatile and suitable for a wide range of cargo, including construction equipment, machinery, vehicles, and other oversized or heavy loads. Many drop deck trailers feature ramps or loading assistance systems that make loading and unloading cargo easier. These ramps can be either manual or hydraulically operated.

Drop deck trailers come in various axle configurations, including single-axle and tandem-axle setups, to accommodate weight and load distribution requirements.

Drum Brakes

Drum brakes are a friction-based braking system used in commercial vehicles to slow down or stop the vehicle (Figure D.7). They are called drum brakes because they employ a cylindrical drum-shaped component that rotates with the wheel. When the brakes are engaged, friction is applied to the inner surface of the drum to decelerate the vehicle [7].

FIGURE D.7 An example of the drum brake configuration.

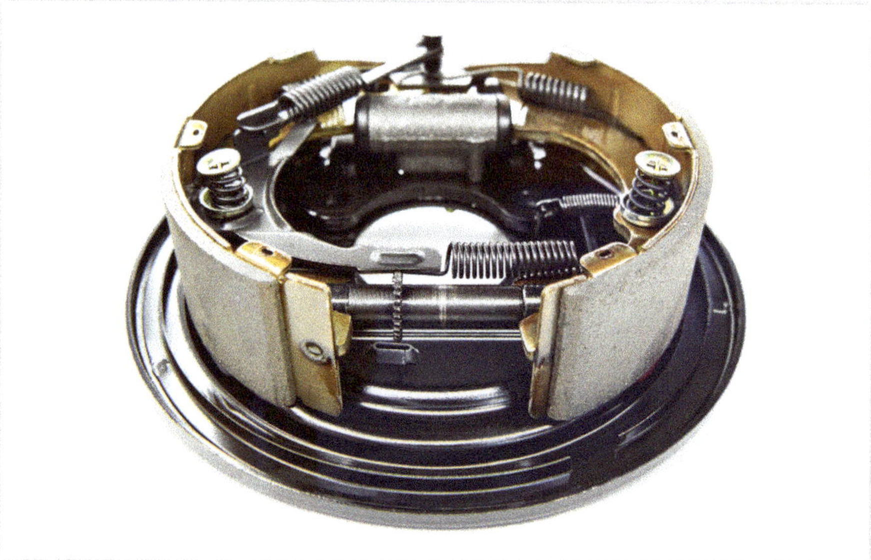

Sergey Ryzhov/Shutterstock.com.

Drum brakes consist of a round or cylindrical brake drum mounted on the axle that rotates with the wheel. The inner surface of the drum is where the braking action occurs. Inside the brake drum, brake shoes are positioned. When the brake pedal is applied, these shoes are pushed outward by hydraulic or mechanical means to contact the drum's inner surface. The brake shoes are equipped with friction linings made of asbestos or more modern alternatives like ceramic or organic compounds. These linings create friction against the drum, generating the braking force.

Commercial vehicles can have either hydraulic or mechanical drum brake systems. Hydraulic systems use brake fluid pressure to activate the brakes, while automated systems rely on linkage and cables. Drum brakes have the advantage of a larger surface area, which helps dissipate heat generated during braking more effectively. This makes them suitable for heavy-duty applications [8].

For more information, see **J1865** [9].

Dry Tank

A dry tank, often called an air tank, is an essential component in the air brake system of commercial vehicles, such as trucks and buses. These tanks are designed to store compressed air and play a critical role in ensuring the proper functioning of the vehicle's pneumatic brake system.

Dry tanks serve as reservoirs for storing compressed air generated by the vehicle's air compressor. This stored air is essential for operating various air brake system components, including brake chambers and control valves. Dry tanks help regulate and maintain the air pressure within the brake system. Proper pressure is crucial for ensuring the effectiveness of the vehicle's brakes.

Compressed air stored in the dry tank is distributed to the brake chambers at the vehicle's wheels. These chambers convert the air pressure into mechanical force to engage the brake shoes or pads. The term "dry tank" reflects its role in moisture removal. Commercial vehicles often have air dryers or desiccant systems connected to the air tanks to eliminate moisture from the compressed air. Dry air is essential for preventing freezing or corrosion within the brake system.

For more information, see **J2806** [8].

Dry Weight

The dry weight of a commercial vehicle refers to its weight without any additional fluids or cargo. It represents the vehicle's base weight, including the chassis, body, engine, and other essential components. Still, it excludes the weight of fuel, fluids (such as oil, coolant, and hydraulic fluids), cargo, passengers, and any optional equipment or accessories. Dry weight is the baseline weight of the commercial vehicle, which serves as a starting point for calculating its overall weight when fully loaded or equipped.

Knowing the dry weight is crucial for determining the vehicle's payload capacity, which is the maximum weight of cargo or passengers it can carry without exceeding its GVWR. Many countries have regulations that specify weight limits for commercial vehicles to ensure safety and proper road infrastructure maintenance. Knowledge of a vehicle's dry weight helps in adhering to these regulations.

Dual Steer

Dual steer refers to a specialized configuration in certain commercial vehicles, primarily trucks, buses, and utility vehicles, with two separate steering systems. This arrangement allows the vehicle to be controlled from either the left-hand side or the right-hand side of the driver's cabin. Dual steer vehicles are often used for specific applications where enhanced visibility, maneuverability, and flexibility are essential.

Dual steer vehicles are equipped with two steering systems—one on the left side and another on the right side of the driver's cabin. This allows the driver to switch between steering positions based on the road layout, traffic conditions, and operational needs. Steering control transfer from one side to the other is typically achieved using mechanical or hydraulic mechanisms that allow the driver to engage and disengage the steering systems as required.

Dual steer configuration enhances maneuverability, especially in urban and congested environments, allowing the driver to quickly navigate tight spaces and negotiate sharp turns. Dual steer vehicles offer improved visibility for the driver when driving on the side of the road where traffic flows, ensuring better awareness of surrounding vehicles, pedestrians, and obstacles.

Dual steer vehicles are commonly used in specialized applications such as waste collection trucks, street sweepers, utility maintenance vehicles, and buses used in countries with varying road traffic rules.

For more information, see **J2425** [10].

Dump Trailer

A dump trailer is a specialized commercial vehicle trailer designed for transporting and unloading loose materials, debris, aggregates, and other bulk cargo. The primary feature of a dump trailer is its hydraulic or mechanical lifting system, which allows the trailer's cargo bed to be raised and tipped, effectively "dumping" its contents onto the ground or at a designated location. Dump trailers are widely used in construction, agriculture, landscaping, and various industries that require efficient and controlled material transportation and unloading.

The cargo bed of a dump trailer is open and usually rectangular, allowing for easy loading and unloading of loose materials. Some models feature removable sides for increased versatility. Dump trailers are equipped with hydraulic or mechanical lifting systems that elevate the front end of the cargo bed. This mechanism allows the bed to be tipped and its contents unloaded.

Dump trailers typically have a hinged tailgate at the rear, which can be opened to release the cargo once the bed is raised. A dump trailer's payload capacity varies based on size, construction, and weight rating. It determines the maximum weight of the cargo that can be safely transported.

Dump trailers often have tandem or tri-axle configurations, distributing the weight of the cargo and providing stability during loading, unloading, and transportation.

Dump Truck

A dump truck is a specialized commercial vehicle for transporting and unloading loose materials, debris, aggregates, and other bulk cargo (Figure D.8). The defining feature of a dump truck is its ability to elevate and tip its cargo bed using a hydraulic or mechanical lifting system, enabling the controlled discharge of its contents. Dump trucks are integral to construction, mining, agriculture, and various industries that require efficient material transportation and unloading.

FIGURE D.8 An example of a typical dump truck, a version of a construction vehicle.

Jonathan Weiss/Shutterstock.com.

The cargo bed of a dump truck is open and usually rectangular, designed for easy loading and unloading of loose materials. Some models feature removable sides for increased flexibility. Dump trucks have a hydraulic or mechanical lifting system beneath the cargo bed. This system raises the front end of the bed, allowing the contents to be tipped out for unloading. Most dump trucks have a hinged tailgate at the rear, which can be opened to facilitate cargo release once the bed is raised.

The payload capacity of a dump truck varies based on factors such as size, construction, weight rating, and axles. It determines the maximum cargo weight that can be safely transported and unloaded. The truck's chassis and suspension systems are engineered to support the weight of the cargo and provide stability during loading, unloading, and transportation. Dump trucks often have tandem or multi-axle configurations to distribute the weight of the cargo and maintain stability during operation.

REFERENCES

1. Society of Automotive Engineers, "J2383 Air Dryer Installation Procedure," SAE Publishing, Warrendale, 2023.
2. Society of Automotive Engineers, "J313 Diesel Fuels," SAE Publishing, Warrendale, 2017.
3. Society of Automotive Engineers, "J2932 Air Disc Brake Actuator Test Requirements," SAE Publishing, Warrendale, 2023.
4. Huang, T., Hu, G., Guo, F., and Zhu, Y., "Investigation of a Model-Based Approach to Estimating Soot Loading Amount in Catalyzed Diesel Particulate Filters," *SAE Int. J. Engines* 12, no. 5 (2019): 567-577, doi:https://doi.org/10.4271/03-12-05-0036.
5. Voutsi, O., Tsinoglou, D., Karamitros, D., and Koltsakis, G., "Pressure Drop of Particulate Filters and Correlation with the Deposited Soot for Heavy-Duty Engines," *SAE Int. J. Adv. & Curr. Prac in Mobility* 2, no. 2 (2020): 692-701, doi:https://doi.org/10.4271/2019-24-0151.
6. Tiwari, A., Durve, A., Barman, J., and Srinivasan, P., "Evaluation of Different Methodologies of Soot Mass Estimation for Optimum Regeneration Interval of Diesel Particulate Filter (DPF)," SAE Technical Paper 2021-26-0208 (2021), doi:https://doi.org/10.4271/2021-26-0208.
7. Society of Automotive Engineers, "J2686 Brake Drum Qualification Recommended Practice," SAE Publishing, Warrendale, 2022.
8. Society of Automotive Engineers, "J2806 Off-Vehicle Brake Testing for Service Brakes Over 10000 Pounds GVW Air, Hydraulic, and Mechanical Actuation," SAE Publishing, Warrendale, 2019.
9. Society of Automotive Engineers, "J1865 Dimensional Compatibility for Commercial Vehicles Fitment of Complete Wheels in Drum Brake Applications - Truck and Bus," SAE Publishing, Warrendale, 2020.
10. Society of Automotive Engineers, "J2425 Steering Control Systems - Heavy Trucks," SAE Publishing, Warrendale, 2021.

E

Earthmoving Machines

Earthmoving machines, often simply referred to as earthmovers, are a diverse group of heavy construction equipment designed for various tasks related to moving, excavating, and shaping earth, rock, and other materials. These machines are essential in commercial and industrial construction, mining, and civil engineering projects, where they perform excavation, grading, trenching, and material handling tasks.

Earthmoving machines encompass many pieces of equipment, including bulldozers, excavators, wheel loaders, backhoes, skid steer loaders, graders, and scrapers. Each type of machine is specialized for specific tasks and conditions. Earthmovers are robustly built with reinforced frames, heavy-duty components, and durable materials to withstand the rigors of construction and excavation work.

Excavators, backhoes, and track hoes are equipped with digging arms (booms) and buckets or attachments for digging, trenching, and excavating soil, rock, or other materials. Wheel loaders, skid steer loaders, and front-end loaders are designed for efficiently scooping, lifting, and transporting materials such as soil, gravel, sand, and debris. Motor graders and road graders are used to grade and level the ground, creating smooth and even surfaces for roads, foundations, and other construction projects.

Bulldozers are characterized by a large blade at the front, used for pushing soil, debris, and other materials and for leveling and grading. Some earthmoving machines, such as compactors and rollers, compact soil or pavement materials to achieve the desired soil density and stability.

Many earthmovers can be equipped with various attachments, such as buckets, rippers, hammers, and augers, to perform multiple tasks on the job site. Earthmoving machines are typically equipped with enclosed operator cabs that provide a comfortable and safe environment for the equipment operator. These cabs often feature advanced controls, air conditioning, and ergonomic seating.

Earthmovers come in both tracked and wheeled configurations. Tracked machines offer superior traction in rough terrain, while wheeled machines provide mobility and speed on smoother surfaces. Safety is a priority, and many earthmovers include features like backup cameras, alarms, and visibility enhancements to prevent accidents on the job site. Advanced earthmoving equipment may be equipped with GPS and telematics systems, allowing precise positioning, tracking, and remote machine performance monitoring. Regular maintenance is crucial to ensure the reliable operation of earthmoving machines. This includes lubrication, inspections, and routine servicing of components.

For more information, see **J1939** [1, 2].

ECM (Engine Control Module)

See **ECU**.

ECU (Electronic Control Unit)

An ECU is critical in commercial vehicles and automotive systems. It is a computerized module responsible for monitoring and controlling various functions within the vehicle, including the engine, transmission, brakes, and other vital systems.

ECUs serve as the vehicle's brain, controlling and regulating various systems, such as the engine's fuel injection, transmission shifting, braking, and emissions. ECUs continuously receive data from various sensors throughout the vehicle, including those related to engine performance, emissions, and safety systems. They process these data and make real-time adjustments to optimize vehicle performance, efficiency, and safety.

ECUs are equipped with diagnostic capabilities to detect and report malfunctions or issues within the vehicle's systems, which can be crucial for maintenance and troubleshooting. Commercial vehicle operators can sometimes reprogram or customize certain aspects of the ECU to suit specific applications, adjust performance characteristics, or improve fuel efficiency.

EDR (Event Data Recorder)

EDR, commonly known as a black box, is installed in many commercial vehicles, including trucks, buses, and passenger cars. It is designed to record crucial data related to the vehicle's operation in the moments leading up to and during an accident or significant event. EDRs are critical in accident reconstruction, safety analysis, and vehicle diagnostics.

EDRs continuously collect and store data related to the vehicle's operation, including speed, engine revolutions per minute (RPM), brake status, throttle position, steering input, airbag deployment, and seatbelt usage. Some EDRs may also record data from additional sensors like GPS and accelerometers. EDRs are programmed to activate and record data in response to specific events. Common triggers include sudden changes in speed (e.g., rapid deceleration), airbag deployment, and pre-crash conditions. Data are typically recorded a few seconds before and after the triggering event.

EDRs are especially valuable in the event of a crash. They record critical crash data, such as the force of impact, the impact's direction, and the crash's duration. This information is crucial for accidents. After a crash, EDRs store data that can be used to assess post-crash conditions, including whether the vehicle was still in motion, the engine was running, and the status of safety systems like airbags and seatbelts. Accident investigators and authorities use EDR data to reconstruct accidents accurately. This helps determine fault, speed at impact, and whether safety systems function correctly.

Vehicle manufacturers and safety researchers use EDR data to improve vehicle safety. Analyzing EDR data from real-world accidents can lead to design improvements in commercial vehicles and safety systems. EDRs can assist in diagnosing vehicle malfunctions or system failures by recording data related to engine performance and sensor inputs. This can be valuable for maintenance and troubleshooting.

EDR data raise privacy concerns, as they record detailed information about a vehicle's operation. To address this, many jurisdictions have regulations specifying who can access EDR data, under what circumstances, and with the vehicle owner's consent. Using specialized equipment and software, EDR data can be retrieved by authorized parties, such as law enforcement, insurance companies, vehicle manufacturers, and vehicle owners. Data retrieval typically requires the vehicle owner's consent or a legal order.

EDRs in commercial vehicles are subject to various legislation and regulations, which may vary by jurisdiction. These regulations often dictate the type of data recorded, data retention periods, and privacy safeguards.

EGR (Exhaust Gas Recirculation)

EGR is an emission control technology used in commercial vehicles to reduce the production of NOx in exhaust gases. EGR systems work by recirculating a portion of the exhaust gases back into the engine's combustion chambers, diluting the fresh air intake, lowering combustion temperatures, and reducing NOx emissions.

The primary function of EGR systems is to reduce the emission of NOx, a group of harmful pollutants that contribute to air pollution and can have adverse health effects. EGR systems are equipped with a recirculation valve that controls the flow of exhaust gases back into the engine. The valve opens and closes as needed to achieve the desired EGR rate (Figure E.1).

FIGURE E.1 An illustration of the system in which the EGR is included [3].

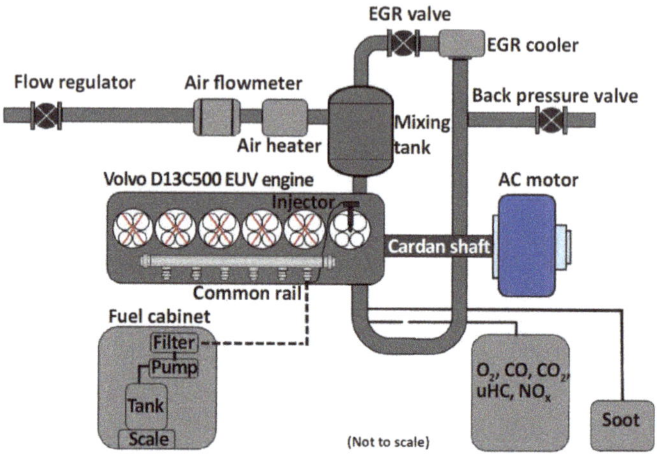

By introducing exhaust gases into the combustion chambers, EGR systems reduce the combustion temperature, lowering the formation of NOx emissions. EGR is an essential technology for commercial vehicles to comply with emission standards and regulations, which are increasingly stringent to reduce environmental impacts.

ELD (Electronic Logging Device)

ELDs electronically record a commercial driver's record of duty status (RODS), ensuring compliance with HOS regulations. ELDs replace traditional paper logbooks. Also, see E-Log Books.

Electric Brake Retarder

An electric brake retarder, also known as an electric retarder or eddy current brake, is a specialized auxiliary braking system used in commercial vehicles to assist with deceleration and reduce wear on the vehicle's primary braking components. It creates resistance through electromagnetic induction, effectively converting the vehicle's kinetic energy into electrical energy, which is dissipated as heat.

Electric brake retarders employ electromagnetic principles to generate resistance against the vehicle's motion. They use the rotation of a magnetic rotor inside a stator to create this resistance, which is directly proportional to the vehicle's speed. The amount of braking force the electric retarder applies can be adjusted by the driver or the vehicle's control system, allowing for fine-tuned deceleration in various driving conditions.

Electric retarders complement the vehicle's primary braking system, such as air brakes or hydraulic brakes, by providing supplementary braking power, particularly during long downhill descents. By reducing the reliance on traditional friction brakes, electric brake retarders extend the life of brake components, decrease maintenance costs, and enhance overall braking system longevity. Some electric retarders can generate electrical energy during deceleration, which can be stored and used to power various vehicle systems or recharge batteries.

Electrical System

The electrical system in commercial vehicles refers to the complex network of electrical components and wiring that provide power, control, and functionality to various systems and components within the vehicle. This system is responsible for powering lighting, starting the engine, managing ECUs, and supporting various auxiliary functions.

One of the primary functions of the electrical system is to start the vehicle's engine using a starter motor, battery, and ignition system. Additionally, the alternator charges the battery and supplies power to other electrical components.

The electrical system powers the vehicle's exterior and interior lighting, including headlights, taillights, turn signals, brake lights, and cabin lighting. Electrical components, such as airbags, ABS, traction control, and ESC, are essential for safety systems.

Commercial vehicles often have electrical components for entertainment and comfort, including radios, heating, ventilation, and air conditioning (HVAC) systems, power windows, and infotainment systems. ECUs are part of the electrical system and manage critical engine functions, such as fuel injection, ignition timing, and emission control. Various auxiliary systems, including power steering, brakes, and liftgates, rely on the electrical system to function.

For more information, see **J2221** and **J1810** [4, 5].

Electrohydraulic Hitch Control

Electrohydraulic hitch control is a sophisticated system used in commercial vehicles, such as trucks and trailers, to manage the operation of hydraulic hitches and coupling mechanisms electronically. This technology provides precise control over the connection and disconnection of trailers, making it safer and more efficient.

Electrohydraulic hitch control combines electronic and hydraulic systems to manage trailer coupling. Electronic components, such as sensors and control modules, provide intelligence and automation. The system utilizes hydraulic power to operate the trailer hitch mechanism. Hydraulic fluid is pressurized and directed to various hydraulic cylinders for precise control.

Many electrohydraulic hitch control systems can be operated remotely, allowing the driver to couple or decouple the trailer from the cab without leaving the driver's seat. This feature enhances safety and convenience. The system often includes sensors and cameras to aid in the precise alignment of the trailer with the hitch, simplifying the coupling process. Electrohydraulic hitch control systems typically have safety features like emergency stop buttons to halt the hitch operation in case of an issue or danger.

Electronic Displacement Control

Electronic displacement control is a technological feature in hydraulic systems used in commercial vehicles, such as trucks and heavy machinery. It provides the ability to electronically control and adjust the displacement of hydraulic pumps or motors, optimizing the flow of hydraulic fluid and enabling precise control over various vehicle functions.

Electronic displacement control integrates electronic components with hydraulic systems, allowing for precisely managing hydraulic pump or motor displacement. This technology enables variable flow rates of hydraulic fluid, which can be adjusted in real time to suit the specific demands of different vehicle functions or equipment attachments. By electronically adjusting the displacement of hydraulic components, commercial vehicles can achieve optimal performance and efficiency in tasks such as lifting, tilting, steering, or any other hydraulic operation.

Electronic displacement control enhances the precision of hydraulic movements, making it easier for operators to perform intricate tasks that require delicate adjustments. The ability to precisely control hydraulic functions contributes to safety, reducing the risk of accidents or equipment damage.

Elevation Mapping

Elevation mapping is a technique that uses GPS technology to create elevation maps of fields to identify areas prone to waterlogging, flooding, or erosion.

E-Log Book

An electronic logbook, an e-log or an ELD, is a digital solution designed to replace traditional paper-based logbooks used by commercial vehicle operators. It provides an automated and accurate method for tracking and recording critical information about a vehicle's operation, driver activities, and compliance with regulations. Electronic logbooks are a cornerstone of modern fleet management technology and are pivotal in enhancing efficiency, safety, and regulatory adherence within the commercial transportation industry.

E-log books automatically capture data such as driving hours, rest periods, engine status, and vehicle movements, eliminating the need for manual entries and reducing the potential for human error. ELDs allow drivers to log in using personalized identification methods such as postal index number (PIN) codes, key fobs, or biometric authentication.

Many electronic logbooks use GPS technology to track vehicle locations, routes taken, and estimated arrival times. ELDs track driving hours, on-duty time, and rest breaks to ensure drivers adhere to HOS regulations set by regulatory authorities like the Federal Motor Carrier Safety Administration (FMCSA) in the US.

ELDs can alert drivers and fleet managers when nearing HOS limits, helping drivers manage their time effectively and avoid violations. Electronic logbooks securely store historical data, allowing for easy retrieval and analysis for regulatory audits, compliance reporting, and operational insights. ELDs are designed to prevent tampering, enhancing the accuracy and integrity of recorded data. Many electronic logbooks can integrate with fleet management software, allowing for streamlined communication, route optimization, and better decision-making.

Engine Displacement

See **Displacement**.

ESC (Electronic Stability Control)

ESC, also known as VSC or electronic stability program (ESP), is an advanced safety system integrated into commercial vehicles. ESC technology enhances vehicle stability and reduces the risk of skidding or losing control in various driving conditions, improving road safety.

ESC systems use various sensors, including wheel speed sensors, steering angle sensors, yaw rate sensors, and accelerometers, to monitor the vehicle's dynamics in real time. When the system detects a loss of vehicle stability or a potential skid, it can automatically assist the driver. This may involve applying individual wheel brakes or adjusting engine power to correct the vehicle's trajectory. ESC often includes traction control functionality, helping to prevent wheel spin during acceleration, especially in adverse road conditions such as rain, snow, or ice.

ESC assists in maintaining proper steering control during emergency maneuvers, reducing the risk of rollovers, and improving the driver's ability to navigate challenging situations. ESC can significantly lower the risk of rollovers, especially in commercial vehicles with a high center of gravity, such as buses and trucks.

For more information, see **J2425** [6], **FMVSS 203**, and **Canadian Motor Vehicle Safety Standard (CMVSR) 203**.

Evaporation E-Viscous Fan

An e-viscous fan, short for electronic viscous fan, is an advanced cooling system component used in commercial vehicles, such as trucks and buses, to regulate engine temperature. Unlike traditional engine cooling fans that rely solely on a mechanical clutch, the e-viscous fan incorporates electronic control to optimize cooling and reduce energy consumption.

E-viscous fans can adjust their rotational speed based on engine cooling needs. Electronic sensors monitor engine temperature, coolant temperature, and other relevant parameters, allowing the fan to operate at variable speeds. By adjusting the fan speed to match the cooling requirements, e-viscous fans improve the overall efficiency of the cooling system. They consume less power than fixed-speed fans, saving energy and reducing fuel consumption. The variable speed control feature reduces noise levels compared to constant-speed engine fans. This contributes to a quieter and more comfortable driving experience. E-viscous fans can respond to rapid changes in engine temperature, ensuring the engine remains within the optimal operating temperature range. This feature is precious during heavy loads or high-temperature conditions. Some e-viscous fans are equipped with remote control and diagnostic capabilities, allowing operators or mechanics to monitor and adjust the fan's behavior for optimal cooling performance.

Excavator

An excavator is a heavy-duty construction machine used for various industries' digging, excavating, and earthmoving tasks. These vehicles, also known as diggers, feature a rotating platform (house) mounted on an undercarriage with a boom, arm, and bucket attachment. Excavators are versatile and widely used for tasks ranging from digging foundations to demolition and material handling (Figure E.2).

The upper portion of an excavator, known as the house or superstructure, can rotate 360° on the undercarriage. This rotation provides flexibility in positioning the boom and bucket. Excavators consist of a boom (extending vertically), an arm (articulated arm connecting the boom to the bucket), and a bucket attachment. These components allow for precise digging, lifting, and material handling. The undercarriage supports the entire excavator and provides mobility. It typically includes tracks or wheels, sprockets, rollers, and idlers. Excavators are powered by hydraulic systems that control the boom, arm, and bucket movements. Hydraulic cylinders provide the force for digging, lifting, and other functions. The operator's cabin on the rotating platform provides a comfortable workspace equipped with controls, instrumentation, and visibility aids.

FIGURE E.2 Example of excavator controls and functions [7].

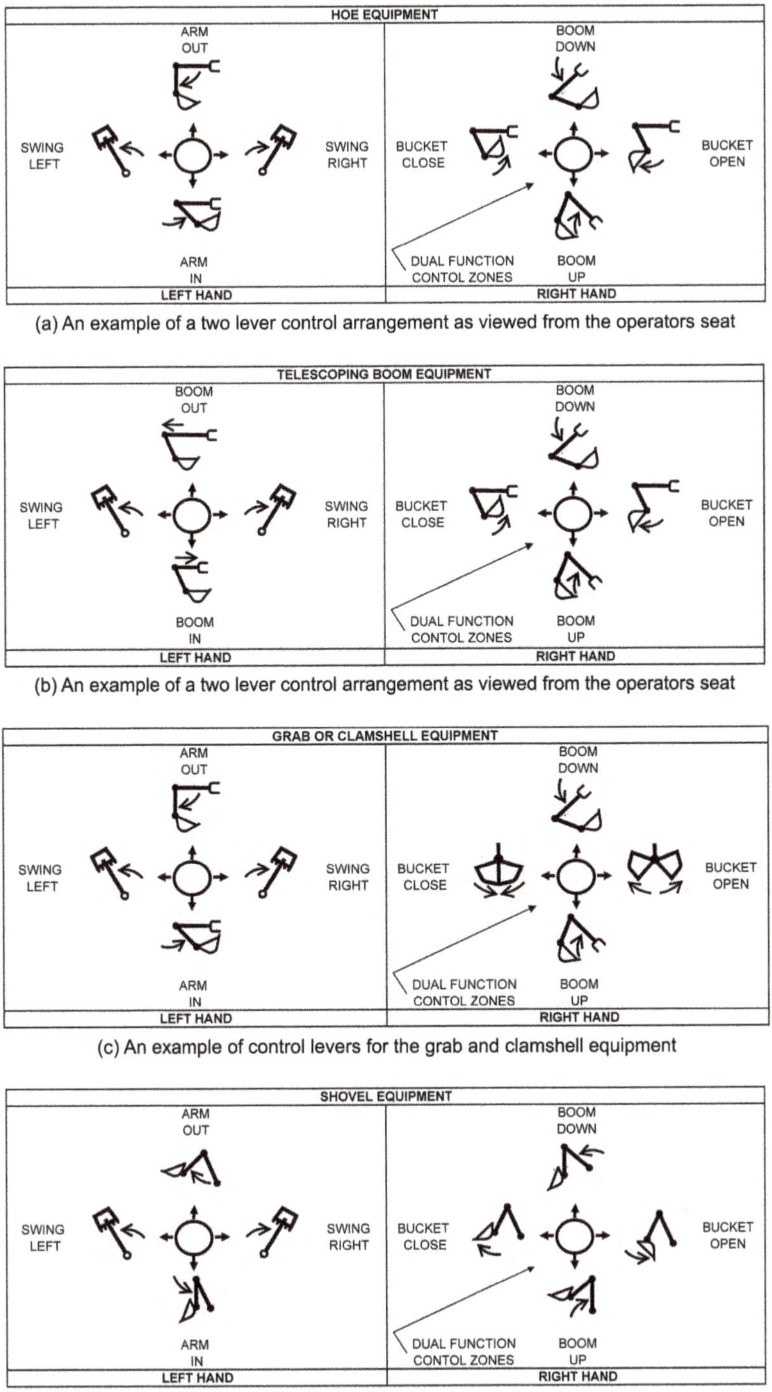

(a) An example of a two lever control arrangement as viewed from the operators seat

(b) An example of a two lever control arrangement as viewed from the operators seat

(c) An example of control levers for the grab and clamshell equipment

(d) Primary control lever functions for right and left hands

REFERENCES

1. Society of Automotive Engineers, "J1939-71 Vehicle Application Layer," SAE Publishing, Warrendale, 2020.
2. Vaidya, V., "Design of Common Electronic Control Unit for Meeting Both 12 Volt and 24 Volt Vehicle Battery Requirements," SAE Publishing, Warrendale, 2020.
3. Goyal, H., Dyuisenakhmetov, A., Ben Houidi, M., Johansson, B. et al., "The Effect of Engine Speed, Exhaust Gas Recirculation, and Compression Ratio on Isobaric Combustion," *SAE Int. J. Engines* 13, no. 5 (2020): 603-615, doi:https://doi.org/10.4271/03-13-05-0038.
4. Society of Automotive Engineers, "J2221 Standardized Symbols for Electrical Circuit Diagrams," SAE Publishing, Warrendale, 2019.
5. Society of Automotive Engineers, "J1810 Electrical Indicating System Specification," SAE Publishing, Warrendale, 2015.
6. Society of Automotive Engineers, "J1865 Dimensional Compatibility for Commercial Vehicles Fitment of Complete Wheels in Drum Brake Applications - Truck and Bus," SAE Publishing, Warrendale, 2020.
7. Society of Automotive Engineers, "J1177 Hydraulic Excavator Operator Controls," SAE Publishing, Warrendale, 2019.

F

Fertilizer Spreader

A fertilizer spreader is an agricultural implement designed for the precise and efficient distribution of fertilizers onto fields or crops. Fertilizer spreaders are created to distribute granular or powdered fertilizers evenly across agrarian fields (Figure F.1). Guidelines may specify standards for the application method to ensure uniform coverage and optimal nutrient distribution [1].

FIGURE F.1 Graphical representation of a GPS fertilizing system [1].

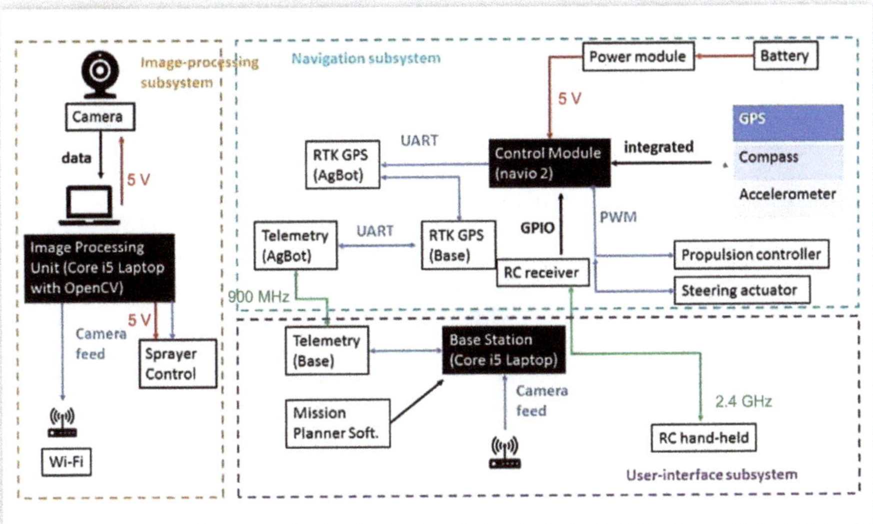

Field Cultivator

Field cultivators are agricultural implements designed for soil preparation and seedbed cultivation. Field cultivators are agricultural implements to prepare the soil for planting by breaking up and aerating the soil, incorporating crop residues, and creating a suitable seedbed. Criteria may be established for the configuration and performance of tillage components, including shanks, sweeps, and rolling baskets. Standards contribute to the effectiveness of soil tillage and residue management.

Field Mapping

Field mapping is a technique that uses GPS to create detailed maps of fields, including topography, soil types, and other relevant information for planting decisions.

Fifth Wheel

A fifth wheel is a vital component of commercial vehicles, especially tractor–trailers or articulated trucks, for coupling the trailer to the tractor unit. This mechanical connection allows for weight transfer and enables the vehicle combination to pivot effectively, providing stability and control during transport [2].

The fifth wheel is a large, flat, plate-like structure mounted on the tractor's chassis. It features a coupling mechanism that securely connects to the kingpin on the trailer. The fifth wheel allows for a pivoting connection between the tractor and the trailer. This pivotal action permits the vehicle combination to turn and navigate corners smoothly. A vital function of the fifth wheel is to transfer the trailer's weight to the tractor unit. This evenly distributes the load between the vehicle's axles, ensuring stability and proper weight distribution for safe driving. Fifth wheels typically have a locking mechanism that secures the trailer's kingpin during transport, preventing accidental uncoupling. Many fifth wheels are adjustable, allowing operators to alter the height to accommodate trailers with varying heights and ensure a level connection (Figure F.2).

FIGURE F.2 The dimensions of the fifth-wheel kingpin [2].

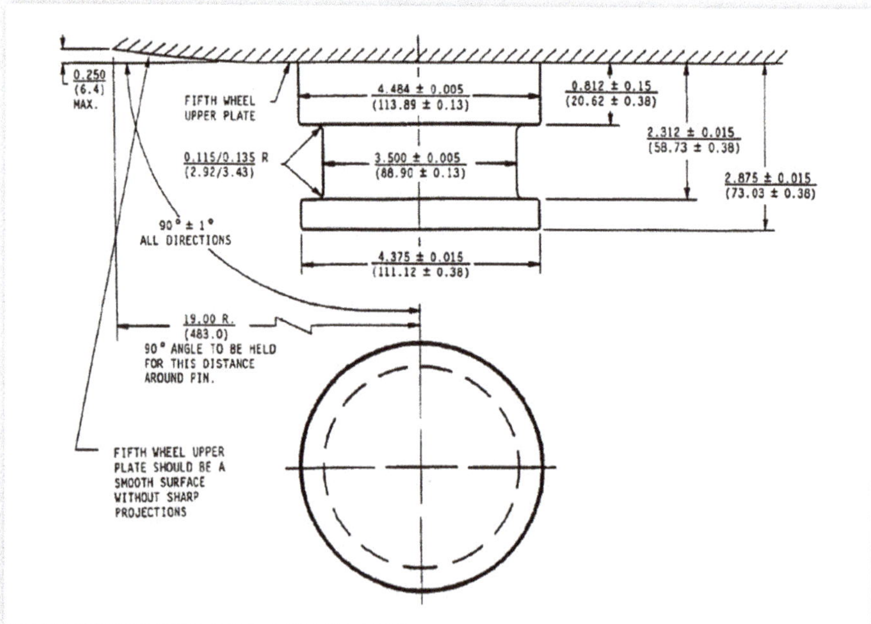

Fire Truck

A fire truck, a fire apparatus or fire engine, is a specialized commercial vehicle designed for firefighting and rescue operations. These vehicles have various tools, equipment, and systems to extinguish fires, rescue individuals from hazardous situations, and provide emergency medical assistance. Fire trucks are critical in ensuring public safety and minimizing property damage during fire emergencies.

Fire trucks have water pumps and tanks supplying pressurized water to hoses, nozzles, and other firefighting equipment. Fire trucks feature hose reels with different types of nozzles for delivering water or firefighting agents to the fire scene. Many fire trucks are equipped with ladders or aerial devices that can be extended to reach the upper floors of buildings for rescue operations and fire suppression. Fire trucks have specialized compartments to store firefighting gear, tools, medical supplies, hydraulic rescue tools (Jaws of Life), and other equipment. Fire trucks are equipped with emergency lights, strobes, and sirens to alert traffic and pedestrians and ensure safe passage to the scene. The cabin of a fire truck accommodates the driver and firefighting crew, providing seating, communication systems, and controls for firefighting operations.

For more information, see **J1212** [3].

Flail Mower

Flail mowers are specialized cutting attachments or implements used in commercial vehicles, such as tractors, agricultural machinery, and landscaping equipment. They are designed to efficiently slice and shred vegetation, including grass, weeds, and small trees, for various purposes, including landscaping, agriculture, and land maintenance.

Flail mowers are equipped with multiple small blades or flails, typically made of metal, which rotate horizontally [4]. These flails are designed to strike and cut through vegetation. They are versatile and can slice a wide range of vegetation, from grass and weeds to light brush and saplings, making them suitable for various applications, including roadside maintenance, pasture management, and landscaping. Flail mowers are known for their durability, solid construction, and robust components to withstand the rigors of cutting through tough and dense vegetation. Many flail mowers include safety features such as guards or shields to contain flying debris and protect the operator and surrounding areas from potential hazards. Some flail mowers allow operators to adjust the cutting height, enabling precise control over the desired vegetation height.

Flatbed Trailer

A flatbed trailer is a commercial vehicle trailer with a flat, open platform without sides or a roof. It is designed to transport various goods and cargo, offering a versatile and accessible hauling solution for the construction and logistics industries.

Flatbed trailers are indispensable in the transportation and logistics industry, offering a flexible and efficient solution for hauling various types of cargo. Flatbed trailers are widely utilized in construction, manufacturing, and freight industries and provide seamless movement of goods and materials across diverse supply chains.

Flatbed Truck

A flatbed truck is a specialized commercial vehicle with an open cargo-carrying bed or platform without sides or a roof. This design allows for the transportation of various types of cargo, equipment, and materials that may not fit within enclosed trucks or require easy loading and unloading from the sides or rear of the truck. Flatbed trucks are widely used across industries for transporting oversized, heavy, and irregularly shaped loads.

The defining feature of a flatbed truck is its open and level cargo bed, providing a versatile platform for transporting cargo of various sizes and shapes. Flatbed trucks come in different lengths and widths to accommodate varying cargo dimensions, from standard pallets to oversized machinery. Flatbed trucks have multiple tie-down points, stake pockets, and anchor points that allow cargo to be securely fastened using chains, straps, and tensioners. The payload capacity of a flatbed truck depends on its size, weight rating, and construction, allowing for the transportation of significant loads. Flatbed trucks can have different cab configurations, including standard, crew, and sleeper cabs, to accommodate driver needs and passenger transport. The truck's chassis and suspension systems are designed to support the weight of the cargo and provide stability during transport.

Flex Fuel

Flex fuel, short for flexible fuel, refers to a type of alternative fuel used in commercial vehicles that can be blended with varying proportions of gasoline and ethanol. It is designed to provide flexibility in fuel choices and reduce the environmental impact of transportation.

Flex fuel vehicles are engineered to run on E85, a mixture of up to 85% ethanol and 15% gasoline, as well as regular gasoline. This flexibility allows drivers to choose between the two fuels based on availability and cost. The critical component of flex fuel is ethanol, typically produced from renewable sources like corn or sugarcane. The ethanol content can vary from E10 (10%)

to E85, depending on the available fuel blend. Ethanol in flex fuel can reduce greenhouse gas emissions compared to pure gasoline, which is considered a cleaner-burning fuel. Using ethanol in flex fuel reduces petroleum consumption and supports the development of renewable and sustainable energy sources [5].

FMS (Fleet Management System)

FMS involves using software and hardware to monitor, track, and manage a fleet of vehicles. It includes vehicle tracking, route optimization, maintenance scheduling, fuel management, and driver performance monitoring.

Food Delivery Vehicle

A food delivery vehicle is a specialized commercial vehicle designed for transporting and delivering prepared food items from restaurants, eateries, and food service establishments to customers' locations (Figure F.3). These vehicles play a crucial role in the modern food delivery industry, ensuring timely and efficient delivery of meals, and enhancing customer convenience.

Food delivery vehicles often have insulated compartments or storage areas that help maintain the temperature and quality of hot and cold food items during transit. Vehicles may have shelves, racks, or secure compartments to organize and prevent food items from shifting during transport. Some food delivery vehicles have heating and cooling systems to ensure food remains at the appropriate temperature throughout delivery. Many food delivery vehicles have GPS navigation systems to help drivers locate delivery addresses and optimize routes. The driver's cabin is designed for comfort and ease of operation, with adjustable seating, user-friendly controls, and visibility aids. Delivery vehicles often feature branding, logos, and signage of the food service establishment to promote brand recognition.

FIGURE F.3 There is a range of food delivery vehicles.

Food Truck

A food truck is a mobile commercial vehicle equipped with a kitchen and cooking facilities designed to prepare, cook, and serve a variety of food items to customers on the go. Food trucks are a popular and dynamic segment of the food service industry, offering various cuisines and culinary experiences in multiple locations.

The interior of a food truck is designed as a fully functional kitchen, complete with cooking equipment, food preparation surfaces, storage, and serving areas. Food trucks have cooking appliances such as grills, stoves, fryers, ovens, and other specialized equipment tailored to the cuisine served. Food trucks have storage compartments and refrigeration units to store ingredients, perishable items, and prepared foods at appropriate temperatures. The exterior of the food truck features one or more serving windows through which customers can place orders and receive their food. Many food trucks have generators or power sources to provide electricity for cooking, refrigeration, lighting, and other operational needs.

Forage Harvester

Forage harvesters, silage harvesters, or choppers are specialized agricultural machines commonly used in commercial vehicles and farming equipment. They are designed to harvest and process forage crops, such as grass, corn, and other plants, to create silage, a valuable livestock feed and forage storage option.

Forage harvesters are equipped with cutting and chopping mechanisms, often as rotating blades or rollers. They efficiently cut and reduce the harvested crops into small, manageable pieces. These machines typically incorporate gathering components, like headers or feed rolls, to collect the standing crop and feed it into the cutting mechanism for processing. The primary function of forage harvesters is to create silage. The chopped forage is deposited into a wagon or silo, where it undergoes controlled fermentation. Silage serves as a valuable feed source for livestock, preserving the nutritional value of forage crops. Many forage harvesters allow operators to adjust the cutting length, enabling customization of the size of the chopped forage for specific livestock and feeding requirements. Depending on their design and size, forage harvesters are powered by various sources, including tractors, self-propelled engines, or power take-off (PTO) systems.

Forestry Machine

A forestry machine, or a forestry vehicle, is a specialized commercial vehicle designed for the unique requirements of forestry and logging operations (Figure F.4). These machines are crucial in efficiently harvesting, processing, and transporting timber and other forest products.

Some common types of forestry machinery are harvesters specialized for tree felling, delimbing, and cutting operations; harvesters efficiently process trees into logs. Another type is the forwarder, designed for transporting logs from the forest to collection points; forwarders often feature load-carrying capacities and articulated steering for maneuverability. Then, skidders drag logs from the cutting site to accessible areas, utilizing winches or grapples for efficient log retrieval. Lastly, mulchers are designed to clear vegetation and small trees, improving access to logging sites.

FIGURE F.4 A drawing of the layout of a wheel-mounted forwarder [6].

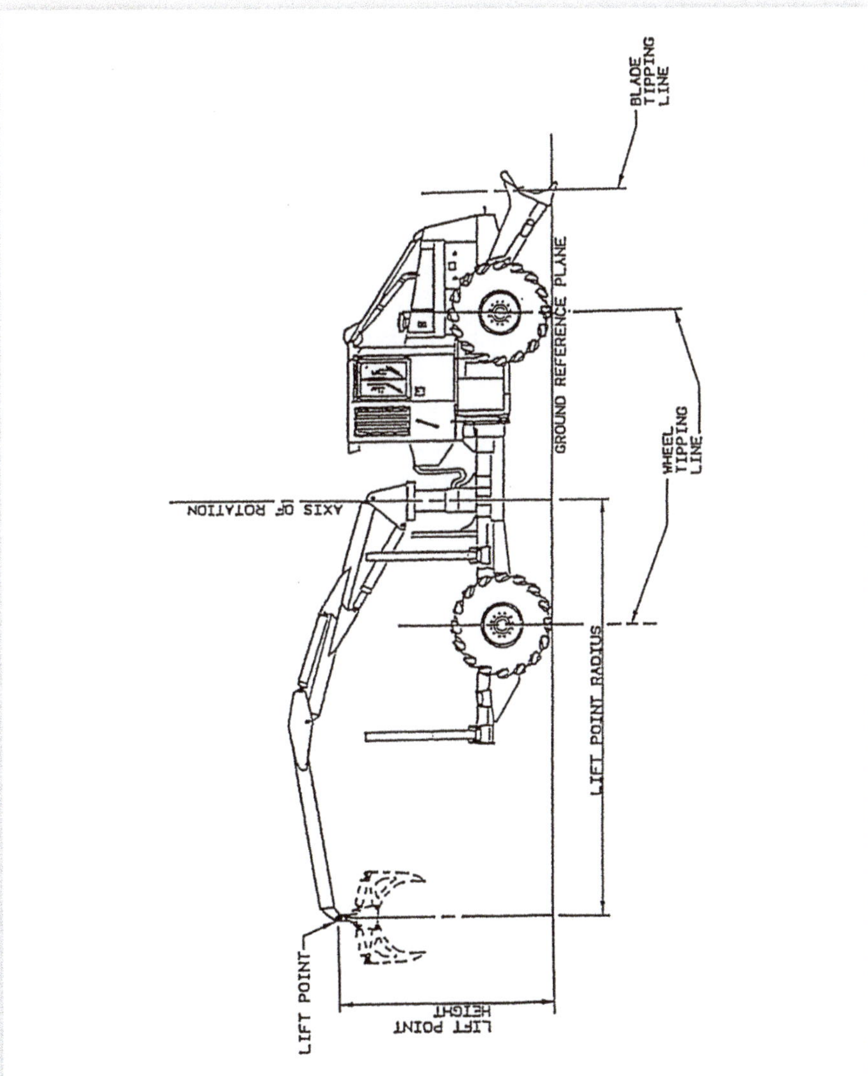

Forklift

A forklift, a lift truck, fork truck, or powered industrial truck, is a specialized commercial vehicle designed for lifting, moving, and transporting heavy loads in warehouses, manufacturing facilities, construction sites, and other industrial environments. Forklifts are critical in material handling, providing efficient and controlled movement of goods, pallets, and containers.

The primary feature of a forklift is its lifting mechanism, which consists of two vertical forks that can be raised and lowered using hydraulic systems. Depending on the specific model and intended use, forklifts are available in various load capacities, ranging from a few hundred kilograms to several tons. The mast is the vertical assembly on the forklift that holds the forks and allows them to be raised or lowered. The mast may have single or multiple stages for increased lifting height. Forklifts are operated using controls in the operator's cabin or standing platform. These controls allow for precise movement, lifting, and tilting of loads. Forklifts have various tires, including solid, pneumatic, and cushion tires, suited to different surfaces and operating conditions. Forklifts have an offset at the rear to balance the weight of the lifted load and prevent the vehicle from tipping over.

Frac Truck (Hydraulic Fracturing)

Frac truck, short for hydraulic fracturing truck, refers to a specialized heavy-duty vehicle used in the oil and gas industry for hydraulic fracturing, also known as "fracking." Frac trucks are essential components of the hydraulic fracturing process, which involves injecting fluid and proppants into underground formations to enhance oil and gas extraction.

Frac trucks have high-pressure pumping systems that generate the pressure to inject fluids, chemicals, and proppants into the wellbore. These trucks have storage tanks for the fracturing fluids, including water, chemicals, and proppants. Frac trucks often have mixing units that blend water, chemicals, and proppants to create the fracturing fluid mixture. The pumping equipment on frac trucks delivers the fracturing fluid mixture at high pressures into the well to create fractures in the underground rock formations. Frac trucks have sophisticated control systems that allow operators to monitor and adjust the pumping process in real time.

Frame Design

The material used for constructing a frame will differ based on the operating characteristics of the truck. Materials can range from medium carbon steel, medium carbon heat-treated steel, chrome manganese heat-treated steel, and chrome manganese molybdenum heat-treated steel, or, for lightweight applications, an aluminum alloy.

Front-End Loader

Front-end loaders are versatile heavy equipment machines designed for material handling and loading tasks in construction, agriculture, and various industrial applications. SAE International establishes standards and terminology to ensure clarity and consistency within the automotive industry, particularly concerning these commercial vehicles' design, performance, and safety aspects. Front-end loaders are equipped with a front-mounted bucket or scoop attachment designed for lifting, loading, and transporting various materials, such as soil, gravel, sand, or other loose materials.

Fuel Management System

Fuel management systems monitor fuel consumption, track fuel expenses, and identify fuel efficiency trends. They help optimize fuel usage and reduce costs.

For more information, see **J703** [7].

Fuel Truck

A fuel truck, also known as a fuel tanker truck or refueler truck, is a specialized commercial vehicle designed for transporting and delivering fuel, such as gasoline, diesel, jet fuel, or other petroleum products, to various locations. These trucks play a critical role in supplying fuel to gas stations, airports, construction sites, industrial facilities, and other places where a consistent and reliable fuel supply is essential (Figure F.5).

FIGURE F.5 Fuel truck delivery of fuel to a fuel station.

Chatchawal Phumkaew/Shutterstock.com.

The defining feature of a fuel truck is its tanker trailer, which is specifically designed to transport large quantities of fuel safely and efficiently. Fuel trucks often have multiple compartments within the tanker to haul different types of fuels simultaneously, reducing the need for separate trips. Fuel trucks have a pumping system to load fuel at refineries or distribution terminals and then unload it into storage tanks or vehicles at the destination. These trucks have precise metering and control systems that ensure accurate delivery of fuel quantities, often integrated with electronic displays and controls in the driver's cabin.

Fuel trucks have safety features such as grounding equipment, overfill protection, and emergency shut-off systems to prevent fuel spills and accidents. Due to the nature of the cargo, fuel trucks must comply with strict regulations and safety standards set by relevant authorities to ensure safe transportation of hazardous materials.

FWD (Front-Wheel Drive)

FWD is a drivetrain configuration in which the power from the engine is primarily transmitted to the front wheels of a propulsion vehicle. SAE International establishes standards and terminology to ensure clarity and consistency within the automotive industry, particularly concerning the design, performance, and safety aspects of vehicles utilizing front-wheel drive.

REFERENCES

1. Khan, N., Medlock, G., Graves, S., and Anwar, S., "GPS Guided Autonomous Navigation of a Small Agricultural Robot with Automated Fertilizing System," SAE Publishing, Warrendale, 2018.
2. Society of Automotive Engineers, "J848 Fifth Wheel Kingpin, Heavy Duty - Commercial Trailers and Semitrailers," SAE Publishing, Warrendale, 2018.
3. Society of Automotive Engineers, "J1212 Fire Prevention on Forestry Machinery," SAE Publishing, Warrendale, 2020.
4. Society of Automotive Engineers, "J1001 Industrial Flail Mowers and Power Rakes," SAE Publishing, Warrendale, 2014.
5. Cohn, D. and Bromberg, L., "Flex Fuel Gasoline-Alcohol Engine for Near Zero Emissions Plug-In Hybrid Long-Haul Trucks," SAE Publishing, Warrendale, 2019.
6. Society of Automotive Engineers, "J2417 Lift Capacity Calculation Method Knuckle-Boom Log Loaders and Certain Forestry Equipment," SAE Publishing, Warrendale, 2020.
7. Society of Automotive Engineers, "J703 Fuel Systems - Truck and Truck Tractors," SAE Publishing, Warrendale, 2015.

G

Garbage Truck

A garbage truck, refuse truck, waste truck, or trash truck is a specialized commercial vehicle for collecting, transporting, and disposing of solid waste and refuse from residential, commercial, and industrial areas. These trucks are crucial in waste management systems, ensuring proper sanitation, cleanliness, and environmental responsibility.

Garbage trucks are equipped with various collection mechanisms, including automated arms, front loaders, rear loaders, and side loaders, designed to lift and load waste containers into the truck's storage compartments. The body of a garbage truck features storage compartments where collected waste is stored until it can be transported to a disposal facility. Many garbage trucks have a compaction system that compresses the waste within the storage compartment, allowing the truck to accommodate more waste before needing to unload. Hydraulic systems control the operation of the collection mechanism, compaction system, and other moving parts of the garbage truck. Garbage trucks are equipped with safety features such as warning lights, backup alarms, and cameras to ensure the safety of workers and pedestrians during waste collection.

GAWR (Gross Axle Weight Rating)

The GAWR is a critical safety specification that applies to each commercial vehicle axle. It represents the maximum permissible weight that an axle is rated to carry, including the weight of the axle itself, its associated wheels and tires, and the load imposed on it.

The GAWR is established to ensure that commercial vehicles operate safely by preventing overloading on individual axles, which could damage the vehicle, compromise handling, and create potential safety hazards. Commercial vehicle manufacturers calculate GAWR based on a combination of factors, including the strength and design of the axle, suspension components, wheels, tires, and other related systems. GAWR information is often provided on a sticker or placard affixed to the vehicle. This label can typically be found on the driver's side door jamb, near the driver's seat, or in the vehicle's owner's manual. Ensuring that the load on each axle remains below its GAWR is vital for proper weight distribution and vehicle stability. Correct weight distribution contributes to safe and predictable handling and reduces the risk of accidents.

For more information, see **J2807** [1].

GCWR (Gross Combination Weight Rating)

The GCWR is a critical specification for commercial vehicles, particularly those with multiple components, such as tractor–trailers or combination vehicles. It represents the maximum permissible total weight of the vehicle (including its payload) and any trailers or towed vehicles when fully loaded.

For more information, see **J2807** [1].

Gear Ratio

Gear ratio, often referred to simply as ratio, is a fundamental concept in commercial vehicles and automotive engineering. It represents the mathematical relationship between the number of teeth on two meshing gears in a mechanical transmission system. Gear ratios are crucial in determining a vehicle's speed and torque output.

Gear ratios are used to create a mechanical advantage by altering the relationship between the rotational speed (RPM) of the input gear (driver gear) and the output gear (driven gear). The primary function of gear ratios is to either increase speed while decreasing torque or decrease speed while increasing torque, depending on the desired performance. In vehicles with manual or automatic transmissions, gear ratios select the optimal combination of speed and torque for various driving conditions, such as acceleration, cruising, or climbing hills.

Commercial vehicles often have different gear ratios in their axles and differentials to optimize power delivery to the wheels, considering load, terrain, and performance requirements. Gear ratios play a role in determining the efficiency of the transmission system. The ideal gear ratio balances performance and fuel economy.

Gearbox

See **Transmission.**

Geofencing

Geofencing, in the context of commercial vehicles, refers to the technology and practice of defining virtual boundaries or geographic zones within which a vehicle operates. These virtual boundaries are established using GPS or other location-based technologies. Geofencing enables fleet managers and operators to monitor and manage vehicles based on geographical location, allowing for enhanced safety, efficiency, and compliance with predefined rules and policies.

Geofencing involves setting up geographical boundaries, often in circles, polygons, or specific routes, that define areas where a vehicle can operate or trigger particular actions. Once a geofence is established, fleet managers can track the vehicle's movements and receive real-time alerts when the vehicle enters or exits the designated zone. Geofencing systems can send instant notifications to fleet managers or administrators when a vehicle breaches a defined geofence, enabling prompt response to unexpected events or deviations. Geofencing technology can trigger automated actions or events, such as sending alerts to drivers, adjusting route plans, or modifying vehicle settings when entering or leaving a specified area. Geofencing can help monitor driver behavior within certain zones, encouraging compliance with speed limits, safety guidelines, and operational protocols.

Geofencing enhances vehicle security by alerting fleet managers and security personnel if a vehicle leaves a designated area without authorization, reducing the risk of theft or unauthorized use. Geofencing can aid in optimizing routes by ensuring vehicles stay within defined zones, adhere to preferred routes, and avoid areas with traffic congestion or road closures.

Georeferenced Planting

Georeferenced planting is a process where crops are planted at precise GPS-defined locations within a field to maximize yield and minimize resource use.

Glad Hand

A glad hand is a crucial component of a commercial vehicle's air brake system, serving as a standardized coupling device that connects the air brake lines of a tractor–trailer or other combination vehicles. The glad hand allows for the secure and efficient transfer of compressed air between the towing vehicle and the trailer, enabling the effective operation of the trailer's brakes and other pneumatic systems (**Figure G.1**).

The glad hand consists of two halves—one on the towing vehicle and the other on the trailer—designed to interlock and form an airtight connection when pressed together. The glad hand provides a connection point for the pneumatic air brake lines of the towing vehicle and the trailer. It allows the compressed air to flow from the towing vehicle's air brake system to the trailer's brake system. When the glad hands are adequately coupled, they create an airtight seal that ensures efficient air transfer and consistent brake operation. This feature is essential for safe and reliable braking while towing. Glad hands are often color-coded to indicate the appropriate airlines that must be connected. Blue is typically used for service brake lines, and red is utilized for emergency brake lines. Some glad hands feature a locking mechanism or latch that secures the connection, preventing accidental disconnection during transit.

FIGURE G.1 The air connections at the back of the tractor cab to connect to the trailer—glad hands.

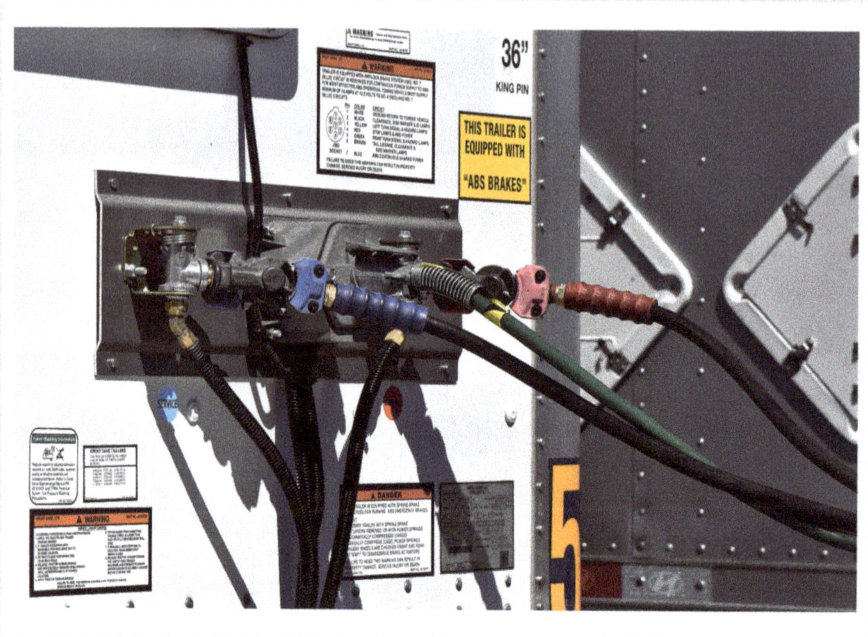

Thomas Trompeter/Shutterstock.com.

Glow Plug

A glow plug is a crucial component in the ignition system of diesel-powered commercial vehicles. It is designed to preheat the air in the engine's combustion chamber, aiding in the ignition of the diesel fuel, particularly during cold starting conditions.

The glow plug features a heating element, usually made of a conductive material, which becomes hot when an electrical current passes through it. This heating element is installed in the combustion chamber of the diesel engine. When the vehicle is started, the glow plug heats the surrounding air in the combustion chamber. This preheating process raises the air temperature to a level that facilitates efficient diesel fuel ignition, particularly in cold weather. Glow plugs ensure that diesel engines start reliably in freezing conditions. Preheating the air reduces the risk of unburned fuel, engine misfires, and excessive exhaust emissions during startup. Glow plugs operate in a duty cycle, which means they are active for a specific period during engine startup and then deactivate once the engine is running at the optimal operating temperature. Many modern commercial vehicles incorporate temperature sensors and control systems to precisely manage the operation of glow plugs, ensuring that they are active for the necessary duration.

GNSS (Global Navigation Satellite System)

A broader term includes GPS and satellite-based navigation systems like Globalnaya Navigazionnaya Sputnikovaya Sistema (GLONASS) and Galileo, often used for precision agriculture.

GPS Guidance

GPS technology is used to provide accurate positioning and guidance for planting equipment.

Grader

A grader, also known as a motor grader or road grader, is a specialized commercial vehicle designed for road construction, maintenance, and grading tasks. Graders are equipped with a large blade that can be adjusted to create a flat, level surface by removing excess material, spreading aggregates, and shaping roadways. These vehicles play a crucial role in achieving smooth and even road surfaces for safe and efficient transportation.

The most distinctive feature of a grader is its adjustable blade, typically positioned between the front and rear axles. The blade can be raised, lowered, angled, and rotated for grading operations. The operator's cabin in a grader is equipped with controls, instrumentation, and visibility aids, allowing the operator to control the blade's movements precisely. Depending on the model, graders may have articulated or frame-steering systems. They are equipped with large, heavy-duty tires designed for traction and stability. Some graders feature rear-mounted scarifiers or rippers that penetrate and loosen compacted soil or material before grading. Graders can be equipped with attachments such as snowplows, spreaders, and compaction rollers to perform additional tasks beyond grading.

For additional information, consult **J1071** [2].

Grain Cart

Grain carts are specialized agricultural implements designed to harvest grain efficiently and transport it from the combine harvester to a waiting transport vehicle or storage location (Figure G.2). SAE International establishes standards and terminology to ensure clarity and consistency within the automotive and agricultural industries, particularly concerning the design, performance, and safety aspects of these commercial agricultural machines.

Grain carts play a vital role in harvesting by assisting combine harvesters in continuous and uninterrupted grain harvesting. They receive and temporarily store harvested grain, allowing the combine to offload without stopping its operation. Grain carts are characterized by their substantial capacity to hold large volumes of harvested grains. SAE International guidelines may specify standards for capacity ratings to ensure efficient and productive harvest operations. Criteria may be established for the design of the grain cart's hitch and towing mechanism. Standards may address factors such as compatibility with different tractor sizes, stability during towing, and ease of hitching and unhitching. Standards may cover specifications for the unloading mechanism, which transfers grain from the cart to a waiting transport vehicle or storage bin. This includes considerations for augers, conveyor belts, or hydraulic systems.

FIGURE G.2 The harvester filling a grain cart.

Solid photos/Shutterstock.com.

Grain Wagon

Grain wagons are specialized agricultural trailers designed for the efficient and bulk transport of harvested grains from the field to storage facilities or processing centers (Figure G.3). SAE International establishes standards and terminology to ensure clarity and consistency within the automotive and agricultural industries, particularly concerning these commercial agricultural trailers' design, performance, and safety aspects.

Grain wagons are characterized by their substantial capacity to carry large volumes of harvested grains, such as wheat, corn, or soybeans. There may be standards for capacity ratings to ensure safe and efficient transport. Criteria may be established for the design and configuration of the trailer, including the number of axles, wheel spacing, and structural integrity. These standards contribute to the stability and load distribution of the grain wagon. Guidelines may encompass specifications for loading and unloading mechanisms, including drills, conveyor belts, or hydraulic systems, ensuring efficient and controlled transfer of grains during loading and unloading operations. Standards may address towing compatibility, including hitch mechanisms and attachment points, if designed to be towed by commercial vehicles. These standards contribute to safe and secure connections between the grain wagon and the towing vehicle.

FIGURE G.3 An example of a grain wagon unloading.

dvande/Shutterstock.com.

Grape Harvester

A commercial grape harvester is a specialized agricultural vehicle designed to automate and efficiently collect grapes from vineyards or grape-producing fields (Figure G.4). These machines typically consist of a chassis mounted with harvesting mechanisms, including mechanical arms, conveyor belts, and shaking systems, which work together to remove grapes from the vines gently. Grape harvesters often have sorting and separation systems to remove leaves, stems, and other debris from the harvested grapes. They are commonly used in large-scale commercial vineyards and wineries to streamline grape harvesting, improve efficiency, and maintain grape quality.

These machines are an essential tool in modern viticulture, as they can significantly reduce the labor required for grape harvesting and ensure a consistent and timely harvest, which is crucial for producing high-quality wines and other grape-derived products.

FIGURE G.4 An example of a grape harvester in the field.

U. J. Alexander/Shutterstock.com.

Grind Soil Sampling

A commercial grind soil sampling vehicle is a specialized agricultural or environmental vehicle equipped with soil sampling equipment to collect soil samples by grinding or drilling into the ground. These vehicles are used in various farming, environmental, and geotechnical applications to gather soil samples for analysis and testing. The sampling process typically involves using rotary augers or drills that penetrate the soil to a predetermined depth, collecting soil samples at various intervals. These samples are then retrieved, stored, and prepared for laboratory analysis to assess soil properties, nutrient levels, contaminants, and other relevant data.

Commercial grind soil sampling vehicles are essential tools for soil scientists, agronomists, environmental consultants, and geotechnical engineers, as they allow for precise and systematic soil sampling, which is crucial for making informed decisions related to land use, crop management, environmental remediation, construction projects, and more. These vehicles may vary in size, design, and sampling capabilities to accommodate different soil sampling requirements and conditions.

GVW (Gross Vehicle Weight)

GVW represents the total weight of a single vehicle (usually the towing vehicle or truck), including its own weight, passengers, cargo, and any additional equipment or accessories, when fully loaded.

GVW focuses solely on a single vehicle and what it can carry, including its payload, passengers, and additional equipment. It does not include any trailers or towed vehicles. GVW is used for single, standalone vehicles to ensure they are not overloaded and remain within the manufacturer's safety and structural limits. It is essential for vehicles like delivery trucks, vans, or buses.

For more information, see **J2807** [1].

GVWR (Gross Vehicle Weight Rating)

The GVWR is a crucial specification for commercial vehicles, indicating the maximum allowable weight that a fully loaded vehicle is rated to support, including its weight, passengers, cargo, and additional equipment. The vehicle manufacturer sets GVWR and plays a pivotal role in maintaining safety and proper weight distribution.

For more information, see **J2807** [1].

REFERENCES

1. Society of Automotive Engineers, "J2807 Performance Requirements for Determining Tow-Vehicle Gross Combination Weight Rating and Trailer Weight Rating," SAE Publishing, Warrendale, 2020.
2. Society of Automotive Engineers, "J1071 Operator Controls for Graders," SAE Publishing, Warrendale, 2014.

H

Haul Truck

A haul truck is a specialized commercial vehicle designed to transport large quantities of materials, such as minerals, ores, aggregates, or waste, within industrial or mining operations. SAE International establishes standards and terminology to ensure clarity and consistency within the automotive industry, particularly concerning these heavy-duty trucks' design, performance, and safety aspects.

HD OBD (Heavy-Duty Onboard Diagnostics)

HD OBD is a sophisticated and standardized system used in commercial vehicles to monitor, diagnose, and report issues related to the vehicle's engine and emission control systems. It is designed to enhance vehicle performance, reduce emissions, and ensure compliance with environmental regulations.

HD OBD continuously monitors the vehicle's engine and emission control systems, collecting data and detecting issues, malfunctions, or deviations from normal operating conditions. When an issue is detected, HD OBD generates diagnostic trouble codes (DTCs) to identify the problem. These codes are standardized and can be read using diagnostic equipment to pinpoint the specific issue. HD OBD plays a crucial role in ensuring commercial vehicles comply with emission standards. It monitors components like EGR systems, DPF, and SCR systems to reduce harmful emissions.

HD OBD provides real-time feedback to vehicle operators and fleet managers, enabling them to address issues promptly, schedule maintenance, and make data-driven decisions to optimize vehicle performance. Commercial vehicles equipped with HD OBD must adhere to regulatory requirements, report emission-related issues, and comply with emission standards. HD OBD can trigger maintenance alerts based on the severity of detected issues. These alerts help prioritize service and repairs (Figure H.1).

For more information, see **J1979** and **J1939** [1, 2].

FIGURE H.1 Applicable SAE and ISO standards for OBD [1].

<table>
<tr><th>Applicability</th><th>OSI 7 Layers</th><th colspan="2">Emissions-Related OBD Communication Requirements</th><th colspan="3">Emissions-related OBD Communication Requirements Based on UDS</th></tr>
<tr><td rowspan="7">Seven layer according to ISO/IEC 7498-1 and ISO/IEC 10731</td><td>Application (layer 7)</td><td colspan="2">ISO 15031-5, SAE J1979</td><td colspan="3">ISO 14229-1 SAE J1979-2</td></tr>
<tr><td>Presentation (layer 6)</td><td colspan="2">SAE J1930DA, SAE J1979DA, SAE J2012DA</td><td colspan="3">SAE J1930DA, SAE J1979DA, SAE J2012DA, SAE J1939DA</td></tr>
<tr><td>Session (layer 5)</td><td colspan="2">ISO 14229-2</td><td colspan="3">ISO 14229-2</td></tr>
<tr><td>Transport (layer 4)</td><td rowspan="2">ISO 15765-2</td><td rowspan="4">ISO 15765-4</td><td rowspan="2">ISO 15765-2</td><td rowspan="4">ISO 15765-4</td><td rowspan="2">ISO 13400-2</td></tr>
<tr><td>Network (layer 3)</td></tr>
<tr><td>Data link (layer 2)</td><td rowspan="2">ISO 11898-1, -2</td><td rowspan="2">ISO 11898-1, -2</td><td rowspan="2">ISO 13400-3</td></tr>
<tr><td>Physical (layer 1)</td></tr>
</table>

Reprinted with permission from F. L. Zarpelon, L. Balcewicz, L. Formolo and R. Ricardo, "Aerodynamic Analysis of a Concept Truck and Trailer for Grain Transportation."

Headliner

The interior roof of any truck.

Heat-Treated Truck Frame

Heat-treated frame rails are identified by a stencil mark on the inside middle section of the rail or a stencil mark on one of the crossmembers, although many newer truck frames have a stencil or sticker attached to the outside of the longitudinal frame rail. When a stencil is used, it shows that the rail is heat-treated. Therefore, the rail flanges must not be drilled or welded since the heat treatment will be affected in that area. Doing so leads to stresses, which are followed by cracking of the frame. Some manufacturers specify in their truck service manuals that only certain strength frame materials should not be drilled. These are generally frames that are manufactured from chrome manganese molybdenum heat-treated steel, which has a maximum safe stress limit of 110,000 psi (758,450 kPa). Some heat-treated frame rails have small patches covering a Brinell hardness test mark along the inside web of the rail. These patches are spaced at 3- to 4-ft (1- to 1.3-m) intervals. The patches can be removed to expose the Brinell hardness markings. Other frame rails may have an "H" stamped on the upper face of the rail flange at about 3 in. (7.62 cm) from the rail end to denote that they are heat-treated frames.

HEUI (Hydraulically Actuated Electronic Unit Injection)

In 1993, Caterpillar and International Truck and Engine Corporation introduced the HEUI system. These injectors are activated by an engine-driven high-pressure pump using engine lubricating oil. Most heavy diesel engines use the engine camshaft to operate the injectors. The pressure of the engine oil, that is, 800 to 3,300 psi, controls the injection rate at up to 30,000 psi, while the amount of fuel injected is determined by the engine management computer using pressure up to 30,000 psi. The HEUI injector solenoid is energized by an electronic signal produced in the ECM. Using input from the sensors, the ECM operates this injector. The variable displacement axial high-pressure oil pump features a built-in reservoir to immediately supply oil at cold starts. The injector actuation pressure control valve is an electronically operated valve that controls the oil pump output and injection pressure.

HOS (Hours of Service)

HOS refers to regulations and guidelines imposed by government authorities, such as the FMCSA in the US, that dictate the maximum time commercial vehicle drivers can spend operating their vehicles and require mandatory rest periods. HOS regulations promote safety by preventing driver fatigue and ensuring drivers have adequate rest and sleep, which is crucial for safe and responsible driving.

HOS regulations specify the maximum number of consecutive hours a commercial driver can drive before taking a mandatory rest break. Drivers must take rest breaks within specific time intervals, ensuring they have opportunities to rest and recharge. HOS rules define the maximum hours a driver can drive within a day. HOS regulations establish the maximum number of driving hours a driver can accumulate within a week. HOS rules mandate specific periods of rest and sleep, including a minimum number of hours spent in a sleeper berth or off-duty.

Hot Oil Truck

A hot oil truck refers to a specialized commercial vehicle used in the oil and gas industry to heat and transfer. Hot oil trucks are designed to transport and circulate heated fluids, such as oil, water, or chemicals, to maintain or restore optimal operating conditions in various oilfield equipment and processes.

Hot oil trucks have a heating system that uses a burner or other heating elements to heat fluids to specific temperatures. These trucks have storage tanks to hold heat-sensitive fluids, including crude oil, water, or various chemicals used in oilfield processes. Hot oil trucks have pumping equipment that circulates the heated fluid from the storage tanks to the equipment or wellsite that requires heating. Many hot oil trucks have temperature control systems that allow operators to regulate and maintain the desired fluid temperature. These trucks have hoses, valves, and connections to transfer heated fluids to and from wellsite equipment.

HP (Horsepower)

HP is a unit of power used to measure the rate at which work is done, or energy is generated within the engine of commercial vehicles. It is a fundamental metric for assessing the performance and capability of vehicle engines.

Hub and Spoke

Hub and spoke refers to a transportation and logistics model that involves centralizing distribution and transportation activities at a central hub, which serves as a main point for sorting, consolidating, and redistributing goods to various spoke locations. This model optimizes the efficiency of cargo movement by utilizing a network of smaller distribution centers or spokes connected to a more prominent central hub.

The central hub is the primary location for receiving, sorting, and consolidating goods from various sources. Spoke locations are smaller distribution centers or points strategically positioned to serve specific regions, cities, or areas. Goods are transported from multiple sources to the central hub, sorted, and dispatched to the appropriate spoke locations. The hub and spoke model allows for efficient consolidation of shipments, enabling economies of scale and optimizing transportation routes (Figure H.2). The spokes are connected to the hub through regular and efficient transportation routes, often using trucks or other transport modes.

FIGURE H.2 An example of hub and spoke shipping and delivery.

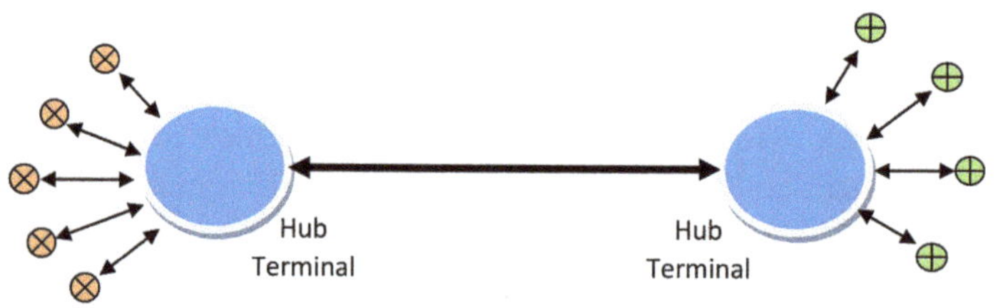

HuckBolt

A HuckBolt, also known as a Huck bolt or structural blind fastener, is a specialized type of fastening and joining method used in commercial vehicles for securely connecting structural components and parts. HuckBolts offer several advantages over traditional fasteners, making them a valuable choice in vehicle assembly and repair.

Hydraulic Brake

A hydraulic brake is a component of the braking system in commercial vehicles. It operates on the principle of hydraulic pressure to slow down or stop the vehicle by applying force to the brake pads or shoes, pressing against the brake discs or drums to create friction and decelerate the vehicle.

Hydromechanical Transmission

A hydromechanical transmission is a type of transmission system used in some commercial vehicles. It combines elements of both hydraulic and mechanical transmissions to provide versatile control over vehicle speed and power delivery.

Hydrostatic Transmission

A hydrostatic transmission, often referred to as a hydrostatic drive or hydrostatic gearbox, is a transmission system used in various commercial vehicles and machinery. Unlike traditional mechanical transmissions, hydrostatic transmissions rely on the principles of hydraulic fluid dynamics to control speed and power distribution.

REFERENCES

1. Society of Automotive Engineers, "J1979 E/E Diagnostic Test Modes: OBDonUDS," SAE Publishing, Warrendale, 2021.
2. Society of Automotive Engineers, "J1939-71 Vehicle Application Layer," SAE Publishing, Warrendale, 2020.

I

I-Beam Steering Axle

The steering axle used for the front axles of most heavy-duty trucks is referred to as an I-beam since it is forged into the shape of an "I." The steering mechanism is connected to steering levers, steering knuckles, and fixed kingpins. All I-beam non-driving front axles are machined from steel or aluminum forgings.

For more information, see **J923_201207**, which outlines the basic nomenclature commonly used for truck and bus drive axle designs [1].

Intermodal

Intermodal refers to a mode of transportation that involves the movement of cargo using multiple modes of transportation, such as trucks, trains, ships, and airplanes. Intermodal transportation aims to optimize the efficiency and convenience of transporting goods by utilizing the strengths of each mode of transport and minimizing the limitations.

Intermodal transportation combines different modes of transport, including road (trucks), rail (trains), sea (ships), and air (airplanes), to move cargo from its origin to its destination. Intermodal systems are designed to seamlessly connect different modes of transport, allowing cargo to be transferred efficiently between them. Freight is typically packed into standardized containers that can be quickly loaded onto trucks, trains, ships, and airplanes without repackaging. Intermodal facilities, such as ports, rail yards, and distribution centers, serve as transshipment points where cargo is transferred from one mode of transport to another. Intermodal transportation leverages the strengths of each mode of transport, optimizing speed, cost, and energy efficiency.

Inventory Management System

In industries with mobile inventories (such as delivery trucks), inventory management systems help monitor stock levels, track goods, and manage replenishment. Inventory management systems can extend to commercial vehicles with inventory in transit.

Inverter

An inverter on commercial vehicles is a device that converts direct current (DC) from the vehicle's electrical system (Figure I.1), often supplied by batteries or the alternator, into alternating current (AC). This allows commercial vehicles to power various electrical devices and appliances that operate on AC power [2].

FIGURE I.1 An example of typical electrical loads that may be found on commercial vehicles [2].

Load	Typical Power Rating
13″ Color TV	70 W
19″ Color TV	300 W
Lap Top Computer	70–90 W
Electric Blanket	150 W
Coffee Maker	1250 W during brewing 200 W during warm cycle
Hair Dryer	800–1875 W
Satellite Receiver	20 W
DVD/VCR	50–100 W
Iron	1200 W
¼″ Drill	300 W
Electric Frying Pan	1500 W
Microwave Oven	1000 W

Irrigation Equipment

Irrigation equipment encompasses a variety of specialized machinery designed to efficiently apply water to agricultural fields, gardens, or landscapes. SAE International establishes standards and terminology to ensure clarity and consistency within the automotive and farming industries, particularly concerning these commercial irrigation systems' design, performance, and safety aspects.

REFERENCES

1. Society of Automotive Engineers, "J923_201207: Nomenclature and Terminology for Truck and Bus Drive," SAE Publishing, Warrendale, 2012.
2. Society of Automotive Engineers, "J2697 Inverter Requirements for Class Eight Trucks - Truck and Bus," SAE Publishing, Warrendale, 2020.

J

J Standards

J Standards, established by SAE International, are a comprehensive set of technical standards and recommended practices used in the automotive industry. These standards cover various aspects, including design, testing, performance, and terminology, to ensure consistency, safety, and interoperability among commercial vehicles and automotive components.

J Standards encompass various specifications and guidelines developed by SAE committees, addressing various aspects of vehicle design, functionality, and performance across multiple categories. SAE J Standards are created with a focus on technical rigor, relying on the expertise of industry professionals, engineers, and researchers to establish benchmarks and best practices.

These standards are interdisciplinary, covering vehicle engineering, safety, materials, electronics, emissions, and more. They are a comprehensive resource for automotive manufacturers, suppliers, and regulatory bodies. SAE J Standards often gain international recognition and adoption, contributing to global harmonization in automotive technologies and practices.

As technology and industry needs evolve, SAE J Standards are periodically updated to incorporate advancements, ensuring that they remain relevant and reflect the latest innovations in the automotive sector. The development of SAE J Standards involves collaboration among industry experts, academia, government bodies, and other stakeholders. This collaborative approach enhances the credibility and applicability of the standards.

J560

J560 Standard, defined by SAE International, refers to a specific technical standard related to electrical connectors for commercial vehicles, particularly those used in trailer connections. These standards are part of the broader collection of SAE J Standards and are crucial for ensuring compatibility and reliability in the electrical systems of commercial trucks and trailers [1].

SAE J560 Standards specifically address the design, configuration, and performance requirements of electrical connectors in the interface between commercial trucks and trailers. The standards aim to establish a common platform for electrical connections, promoting consistency and interoperability.

J560 Standards define various connector types, specifying the number and configuration of pins and ensuring a standardized approach for different electrical functions such as lighting, braking, and power supply between the truck and the trailer (Figure J.1).

FIGURE J.1 The physical layout of the J560 connector [1].

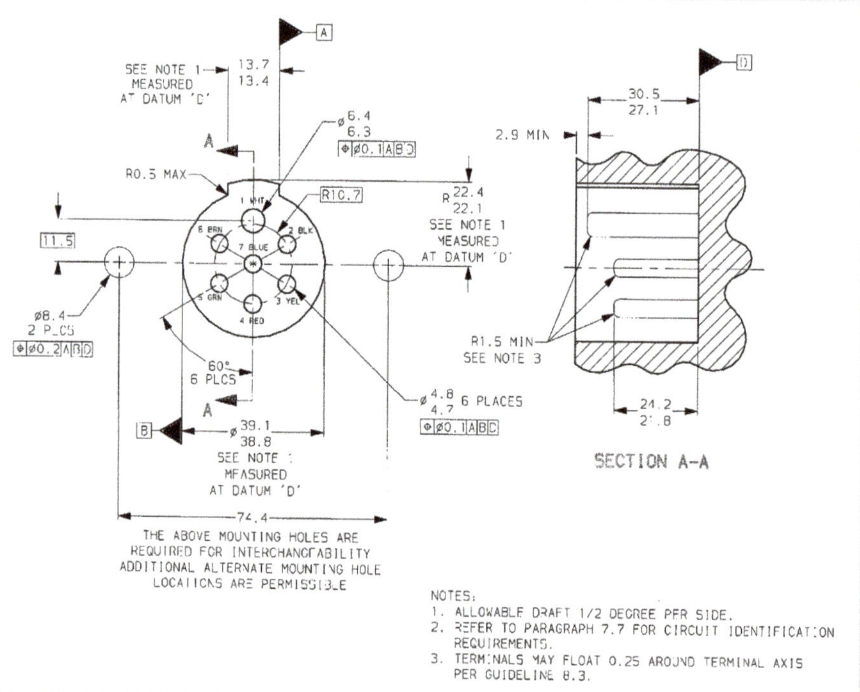

Jack Knife

Jackknife refers to a potentially dangerous situation that can occur with articulated vehicles, such as tractor–trailers or articulated buses. This term describes the folding motion resembling the closing of a pocket knife, where the trailer or towed unit swings outward, leading to a severe misalignment between the tractor and the trailer.

Jackknifing is a phenomenon primarily associated with articulated vehicles, which consist of a towing unit (tractor) and a separate trailer or towed unit. The term is particularly relevant to semi-truck hauling trailers. Jackknifing occurs when the trailer loses traction or experiences a skid, leading it to swing outward, forming an acute angle with the tractor. This loss of control can be triggered by factors such as slippery road conditions, sudden deceleration, or inappropriate braking.

The term jackknife specifically emphasizes the folding motion resembling the action of a pocket knife. This situation can be highly hazardous, as the swinging trailer can potentially collide with the tractor, nearby vehicles, or obstacles, leading to accidents.

Jake Brake

A Jake Brake, officially known as an engine compression release (ECR) brake, is a specialized engine braking system commonly used in commercial vehicles. The Jake Brake allows for controlled and efficient deceleration by modifying the operation of the engine's exhaust valves during the compression stroke.

The Jake Brake operates by momentarily opening the exhaust valves in the cylinders just before the compression stroke. This releases the compressed air in the cylinders, creating a braking effect by utilizing the engine's natural compression resistance. This feature is handy for managing downhill descents or controlling speed without relying solely on conventional friction brakes.

To mitigate noise concerns associated with engine braking, especially in residential or noise-sensitive areas, SAE standards may include criteria for noise reduction features, such as mufflers or sound attenuation systems. Guidelines may encompass specifications for the driver interface, ensuring that the Jake Brake is easily controlled and modulated by the driver, enhancing safety and usability.

Jamb Switch

A jamb switch is an electrical grounding switch that operates the interior lights of a truck when the door is opened. It is located in the door jam.

REFERENCE

1. Society of Automotive Engineers, "J560 Primary and Auxiliary Seven Conductor Electrical Connector for Truck-Trailer Jumper Cable," SAE Publishing, Warrendale, 2020.

K

Kill Truck

A kill truck refers to a specialized commercial vehicle used in the oil and gas industry for well-controlled operations, specifically during well-controlled incidents or interventions. Kill trucks are equipped with equipment and fluids necessary to control pressure and fluids within a wellbore to regain control over the well and ensure safety during drilling, completion, or workover operations.

Kill trucks are equipped with pressure control equipment, including high-pressure pumps and valves, to manipulate the pressure and flow of fluids within the wellbore. These trucks have storage tanks to hold the fluids required for well-controlled operations, such as drilling mud or specialized kill fluids. Kill trucks have pumping systems that deliver fluids at high pressures into the wellbore to counteract any unwanted pressure imbalances. Many kill trucks have pressure monitoring and control systems to ensure accurate safe well-controlled operations.

Kill trucks are primarily used in the oil and gas industry for well-controlled operations. Kill trucks are deployed during well-controlled emergencies, such as blowouts or kicks, to regain control over pressure and fluid flow within the wellbore. Kill trucks may be used during drilling operations to maintain wellbore stability and control pressure while drilling into formations with varying pressures. Kill trucks are used during well workover or completion activities to control and manipulate fluid levels and pressures.

Kingpin

The kingpin connection is a crucial coupling mechanism that links the tractor (the prime mover or towing vehicle) to the trailer in commercial vehicle combinations (Figure K.1). This connection allows the tractor to pull the trailer, enabling articulation for turning and maneuvering.

FIGURE K.1 Dimensions of the upper couple kingpin [1].

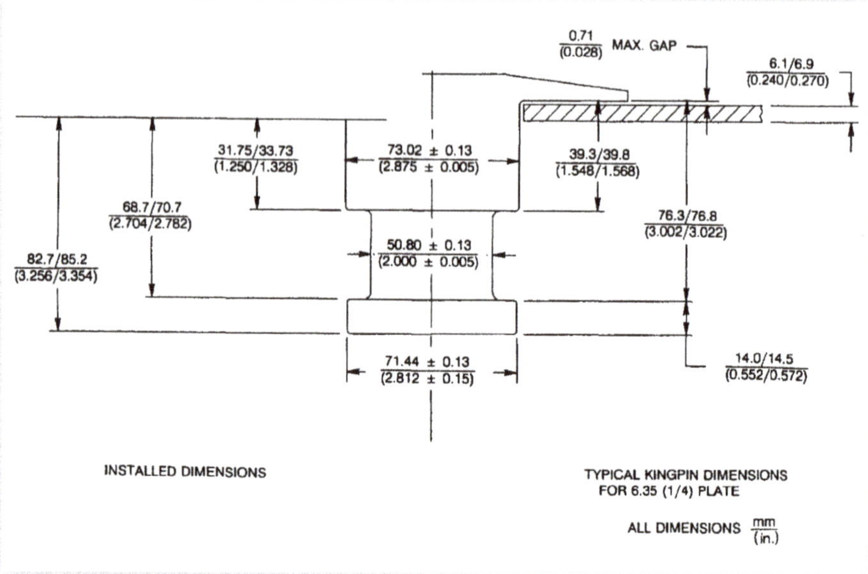

The kingpin serves as the pivot point at the front of the trailer, allowing the trailer to articulate concerning the tractor. This pivotal connection enables the combination to navigate corners and curves. The kingpin connection ensures a secure attachment between the tractor and the trailer, preventing separation during transit and enhancing the safety and stability of the entire combination. The kingpin is responsible for transmitting the vertical and horizontal forces between the tractor and the trailer, including the weight of the trailer and its cargo to the tractor's chassis and suspension. The kingpin connection helps distribute the weight of the trailer and cargo over the axles of the tractor and trailer, ensuring proper balance and adherence to weight regulations.

For additional information, see **J700** and **J848** [1, 2].

KS (Knock Sensor)

KS is a piezoelectric circuit that detects spark knock, pre-ignition, and detonation.

REFERENCES

1. Society of Automotive Engineers, "J700 Upper Coupler Kingpin - Commercial Trailers and Semitrailers," SAE Publishing, Warrendale, 2020.
2. Society of Automotive Engineers, "J848 Fifth Wheel Kingpin, Heavy Duty - Commercial Trailers and Semitrailers," SAE Publishing, Warrendale, 2018.

L

Ladder Frame

Most heavy-duty truck frames are of the parallel ladder design, assembled either by bolting or by riveting. The two main parts of the frame are the two longitudinal members, which are often referred to as "rails." These rails actually carry the load, while the transverse or crossmembers tie the two rails together to provide strength and rigidity, as well as weight distribution over both rails equally. The frame must be stiff enough to support the loads placed upon it during operation and must be capable of absorbing the torque transfer from the axles, which attempt to twist the frame. Thus, the frame must be flexible enough to distribute these stresses throughout its length. The parallel-construction-type frame (chassis) is the most common design, yet many trucks may have a frame known as a drop-center frame because the frame side rails drop in height immediately behind or in front of the support member for the fuel tank. This results in a thicker frame that extends throughout the midsection of the construction. This design results in a lower overall center of gravity and increased strength.

Landing Gear

Landing gear, also known as landing legs or support legs, is an integral part of semi-trailers. It supports the trailer when it is not connected to a tractor, such as during loading, unloading, or stationary (Figure L.1).

FIGURE L.1 An illustration of the trailer landing gear that holds the front of the trailer up during detachment from the tractor.

Kevin Norris/Shutterstock.com.

Leaf Spring

A leaf spring is a suspension component commonly used in commercial vehicles to absorb and distribute the forces experienced during vehicle motion. SAE International establishes standards and terminology to ensure clarity and consistency within the automotive industry, particularly concerning the design, performance, and safety aspects of these essential suspension elements (Figure L.2).

Leaf springs consist of multiple thin, flat, and curved steel plates (leaves) layered on each other. The leaves are secured at the center and gradually decrease in length toward the ends, forming a tapered and flexible assembly. Leaf springs are a critical component of a vehicle's suspension system, supporting the vehicle's weight, absorbing shocks from road irregularities, and contributing to overall ride comfort.

Standards may cover various designs of leaf springs, including multi-leaf and parabolic configurations, each offering different characteristics regarding load distribution, ride quality, and stability.

FIGURE L.2 An illustration of a leaf spring [1]; (a) single spring installation assembly (b) detail of spring end (c) single spring.

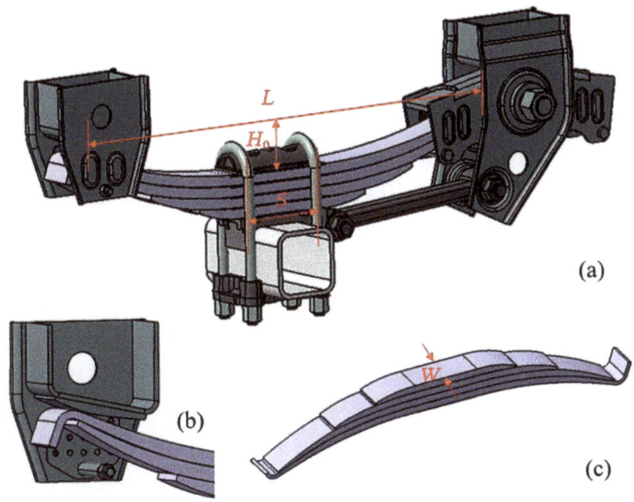

Lift Axle

Lift axles represent a dynamic advancement in vehicle design, offering a versatile solution to optimize load-bearing capacity and improve fuel efficiency in commercial vehicles. These specialized axles can be raised or lowered, allowing vehicles to adapt to varying load sizes and road conditions while minimizing wear and tear on tires and suspension systems.

Lift axles can be electronically or pneumatically controlled to raise or lower the additional axle(s) based on the load weight. This adaptive feature distributes the load evenly across multiple axles, enhancing vehicle stability and reducing tire stress. Lift axles help vehicles meet legal weight regulations by enabling them to carry heavier loads when necessary while maintaining compliance during lighter load periods. By lifting the axles when not needed, vehicles experience reduced rolling resistance, improving fuel efficiency. This is particularly advantageous for long-haul trucking and transportation fleets.

In situations requiring tighter turns or maneuvering in confined spaces, lifted axles enhance vehicle maneuverability by reducing the effective wheelbase, minimizing tire scrubbing, and allowing for smoother cornering. Lift axles enable vehicles to carry varying payloads while remaining within weight limits, providing flexibility in load configurations.

Liftgate

A liftgate is a versatile and indispensable tool in logistics and transportation. It is an innovative hydraulic or mechanical platform installed at the rear of trucks, trailers, or vans. It is designed to effortlessly load and unload a wide range of goods, from heavy equipment to fragile packages. With its ability to raise and lower cargo to ground level, a lift gate simplifies loading and unloading, saving time and effort while enhancing operational efficiency.

Lift gates are available in hydraulic and mechanical configurations, each offering advantages. Hydraulic lift gates use fluid power to raise and lower cargo smoothly, while mechanical lift gates rely on manual operation for lifting. Lift gates are engineered to accommodate various weight capacities, catering to a broad spectrum of cargo sizes and types. This versatility ensures that small parcels and hefty machinery can be easily handled. Lift gates offer different platform sizes and designs, allowing operators to select the ideal platform for their cargo and operational needs. These platforms can be flat, tucked away, folding, or rail-style. Many modern lift gates come equipped with remote control capabilities, enabling operators to control the lifting and lowering of the platform precisely from a safe distance. Lift gates are designed with safety in mind, featuring mechanisms such as non-slip surfaces, safety barriers, and auto-folding ramps to prevent accidents and ensure secure cargo handling.

Light-Duty Truck

Light-duty trucks are commercial vehicles designed for various applications, balancing versatility, fuel efficiency, and payload capacity. SAE International establishes standards and terminology to ensure clarity and consistency within the automotive industry, particularly concerning these versatile trucks' design, performance, and safety aspects [2].

Light-duty trucks are characterized by their ability to handle moderate loads, making them suitable for various applications, including personal use, commercial delivery, and light-duty work in industries such as construction and landscaping. Light-duty trucks, tailored to different applications, emphasize their capability to accommodate various body types, such as pickups, vans, and chassis cabs.

Lime Spreader

A lime spreader, also known as an agricultural lime spreader or lime spreader truck, is a specialized vehicle used in agriculture and related industries to distribute agrarian lime, which is utilized to improve soil quality and adjust pH levels in fields and farmlands.

Live Bottom Trailer

Live bottom trailers are innovative hauling solutions that streamline transporting bulk materials such as aggregates, sand, gravel, asphalt, and agricultural products. These trailers incorporate a unique conveyor belt system that allows for controlled and continuous unloading of materials, eliminating the challenges associated with traditional dump trailers.

The defining feature of live bottom trailers is their integrated conveyor belt system that spans the length of the trailer bed. This system consists of multiple moving slats that push materials out of the trailer, ensuring a steady and consistent flow during unloading. Live bottom trailers handle various materials, from fine-grained substances to larger aggregates. This versatility makes them indispensable in the construction, agriculture, road maintenance, and waste management industries. The continuous unloading process of live bottom trailers minimizes the impact on materials, reducing segregation and preserving the quality of transported goods. This is especially valuable for materials like asphalt and fragile agricultural products.

Operators can precisely control the speed and direction of the conveyor belt, allowing for controlled discharge in tight spaces or when creating an even material spread. Compared to traditional dump trailers, live bottom trailers experience less wear and tear due to their smoother unloading process, reducing maintenance costs and downtime.

Livestock Trailer

Livestock trailers are specialized transport units designed to safely and efficiently transport various types of animals, including cattle, pigs, sheep, and poultry (Figure L.3). These trailers offer a convenient and reliable means of moving livestock from one location to another, whether for farm-to-market transportation, agricultural shows, auctions, or veterinary visits.

Livestock trailers are constructed with sturdy materials and equipped with safety features to ensure the well-being of animals during transit. Robust flooring, ventilation systems, and secure gating mechanisms provide your livestock a safe and comfortable journey. These trailers come in various sizes and configurations to accommodate different types and numbers of animals. From single-deck to multi-level designs, livestock trailers can be customized to suit specific transportation needs. Livestock trailers often feature ramps, chutes, and partition gates that simplify loading and unloading. This minimizes stress on both animals and handlers, resulting in smoother operations. Proper ventilation systems in livestock trailers maintain optimal air circulation, preventing heat stress and ensuring a healthy environment for transporting animals. Livestock trailers are designed to meet regulatory standards for animal transport, guaranteeing that your operations remain compliant with local and national regulations.

FIGURE L.3 An example of a livestock trailer behind a cab over an engine.

Miguel Perfectti/Shutterstock.com.

LNG (Liquid Natural Gas)

LNG is a cryogenic fuel that primarily consists of methane liquefied at extremely low temperatures for use as a clean and alternative fuel in vehicles. Standards and terminology ensure clarity and consistency within the automotive industry, particularly concerning the design, performance, and safety aspects of vehicles utilizing LNG.

Load Planning Software

Load planning software optimizes cargo loading for maximum space utilization and weight distribution, minimizing the trips required.

Load Security

Load security refers to the measures and practices taken to ensure that cargo or freight carried by a commercial vehicle is adequately secured, preventing it from shifting, falling, or causing any hazards during transportation. Load security is crucial for safety on the road, protection of the cargo, and compliance with regulations.

Load security involves using appropriate restraints such as straps, chains, ropes, nets, and tensioning devices to secure the cargo. Distributing the load evenly across the vehicle's axles and ensuring proper weight distribution helps maintain stability and control during transit. Adding blocks, braces, or dunnage between items or around the cargo prevents movement, minimizing the risk of damage. Depending on the type of cargo, equipment such as load bars, ratchets, and anchor points may be used to secure the load. Different jurisdictions have load security regulations, including weight limits, securement methods, and more.

Loader

Loaders are versatile heavy equipment machines designed for material handling and loading tasks in construction, agriculture, and various industrial applications. Standards and terminology are established to ensure clarity and consistency within the automotive industry, particularly concerning these commercial vehicles' design, performance, and safety aspects.

Loaders are equipped with a front-mounted bucket or scoop attachment designed for lifting, loading, and transporting various materials, such as soil, gravel, sand, or other loose materials. Loaders are versatile machines capable of performing tasks beyond simple loading, including grading, digging, and material spreading.

Loading and Unloading Equipment

Loading and unloading equipment refers to various specialized tools, devices, and systems designed to facilitate the efficient loading and unloading of cargo onto and from commercial vehicles. These equipment types are essential for optimizing transportation, ensuring safety, and minimizing the time and effort required to load and unload goods.

Loading and unloading equipment encompasses various devices, including ramps, lift gates, cranes, hoists, and conveyors, each tailored to specific cargo and vehicle configurations. Many loading and unloading equipment can be adjusted to accommodate varying cargo sizes and weights, providing flexibility in the transportation process.

Locking Differential

A locking differential is a type of differential mechanism used in vehicles to enhance traction and improve off-road capabilities. A locking differential can temporarily lock the rotational movement of both drive wheels together, ensuring they rotate at the same speed and providing equal power distribution to both wheels, even in challenging terrain or slippery conditions.

A differential allows the wheels on the same axle to rotate at different speeds, enabling smooth turns on paved roads. However, this can lead to uneven power distribution in certain situations, such as off-road or low-traction conditions. A locking differential can be manually or automatically engaged.

When engaged, it locks the rotation of both drive wheels together, preventing one wheel from spinning freely while the other has no traction. The locked differential ensures that both wheels receive equal power, maximizing traction and preventing one wheel from spinning, while the other remains stationary. Locking differentials are especially useful in off-road environments, where uneven terrain or slippery surfaces can cause wheel spin.

Locking differentials improve traction by ensuring power is sent to both drive wheels, even when one wheel loses traction. Off-road vehicles equipped with locking differentials can navigate challenging terrain more effectively by reducing the likelihood of getting stuck. Locked differentials provide better control in low-traction situations, allowing the driver to maintain direction and stability. Improved traction and control contribute to safer driving, especially in adverse weather conditions or difficult terrains.

Logbook

A logbook, also known as a driver's or vehicle logbook, is an official record that documents essential information related to a commercial vehicle's operation, activities, and maintenance. It is a crucial tool for regulatory compliance, accountability, and effective fleet management in the commercial transportation industry.

The logbook includes details about the driver, such as their name, driver's license number, and contact information. Essential vehicle details, including the vehicle's identification number (VIN), license plate number, make, model, and year, are recorded. Each entry includes the date and time of the vehicle's activities, such as starting a journey, completing a trip, or performing maintenance. For each trip or journey, the log book records the trip's starting and ending locations, mileage, and purpose (e.g., delivery, transportation, personal use). A critical aspect of the log book is recording the driver's HOS, which includes driving, on-duty, off-duty, and rest periods. These records are crucial for compliance with regulations governing driver fatigue and rest breaks. Odometer readings are documented at the beginning and end of each trip to track the distance traveled.

The logbook may include entries related to refueling, maintenance, repairs, and vehicle servicing. This information helps monitor vehicle health and track operating costs. Signatures of the driver and a responsible authority, such as a supervisor or fleet manager, are often required to authenticate the entries in the logbook.

Lowboy Trailer

A lowboy trailer, also known as a lowbed trailer, refers to a specialized type designed to transport heavy and oversized cargo, such as construction equipment, machinery, and other large loads. The defining feature of a lowboy trailer is its low deck height, which allows for transporting tall and heavy items while maintaining a lower center of gravity for stability during transit.

LSD (Limited-Slip Differential)

An LSD, sometimes called a limited-slip diff, is a specialized type of differential used in commercial vehicles (Figure L.4). The primary function of an LSD is to distribute power between the drive wheels while allowing some differentiation in wheel speed, providing enhanced traction and stability, especially in situations where one wheel has reduced grip.

FIGURE L.4 Geared, torque-sensitive limited-slip differential [3].

LSDs are designed to improve traction by allowing power to be delivered to both drive wheels, even when one of the wheels has less traction (e.g., due to slippery or uneven road conditions). This helps prevent wheel spin and loss of control. LSDs distribute torque (power) between the drive wheels based on each wheel's resistance. This ensures that the wheel with grip receives more power while preventing the wheel with less grip from spinning freely. LSDs enhance stability and control, particularly in off-road, slippery, or challenging terrain conditions. They are beneficial for commercial vehicles like trucks and off-road vehicles.

LSDs provide a smoother power transfer between wheels, reducing wheel hop and tire wear while improving overall vehicle handling. LSDs come in various types, including mechanical LSDs, viscous LSDs, and electronic LSDs. Each type functions slightly differently to enhance traction and stability.

REFERENCES

1. Wang, L., Zhu, C., Lu, X., Zhang, Z. et al., "Structural Design and Analysis of Sliding Composite Mono Leaf Spring," *SAE Int. J. Commer. Veh.* 16, no. 3 (2023): 299-312, doi:https://doi.org/10.4271/02-16-03-0020.

2. Society of Automotive Engineers, "J2258 Light Utility Vehicles," SAE Publishing, Warrendale, 2022.

3. Pinho, A. and Franco, C., "Studies on a Usage of a Limited Slip Differential Without Full Locking Features Compared to Open Type Differential and Traction Control in Mid-Size Truck - Drivers for Strategic Decision Balancing Technical Performance and Customer Perception," SAE Technical Paper 2015-36-0164 (2015), doi:https://doi.org/10.4271/2015-36-0164.

M

Maintenance Vehicle

A maintenance vehicle is a specialized commercial vehicle equipped with tools, equipment, and facilities for performing various maintenance and repair tasks. Maintenance vehicles are designed to provide on-site support for servicing other vehicles, infrastructure, or equipment, and they play a crucial role in ensuring the proper functioning and safety of commercial operations.

Manual Transmission

A manual transmission refers to a vehicle transmission system requiring the driver to shift gears using a clutch pedal and gearshift lever manually. In a manual transmission, the driver engages different gear ratios to control the power and torque delivery from the engine to the wheels, allowing for precise control over the vehicle's speed and performance.

A manual transmission is equipped with a clutch system that allows the driver to disengage and engage the connection between the engine and the transmission. The gearshift lever, often located on the vehicle's center console, lets the driver select different gears by moving the lever into specific positions. Manual transmissions offer a range of gear ratios, typically including multiple forward gears and reverse gears, enabling the driver to match the engine's power to different driving conditions. The driver has complete control over gear selection, engaging the clutch to change gears manually based on speed, load, and terrain. To switch gears, the driver depresses the clutch pedal to disengage the engine's power from the transmission, shifts gears using the gearshift lever, and then releases the clutch pedal to engage the new gear.

Manure Spreader

Manure spreaders are specialized agricultural implements designed to efficiently distribute animal waste (manure) onto fields as a nutrient-rich fertilizer (Figure M.1). SAE International establishes standards and terminology to ensure clarity and consistency within the automotive industry, particularly concerning these commercial farming tools' design, performance, and safety aspects.

FIGURE M.1 Manure spreader pulled behind a tractor.

Fotokostic/Shutterstock.com.

Manure spreaders typically feature a loading mechanism, including a conveyor belt or a beater mechanism, to collect and evenly distribute manure across a designated area. A spreading mechanism ensures an even and controlled distribution of manure over the field. This may involve considerations for adjustable spread patterns and distribution rates.

Mechanical Transplanter

Mechanical transplanters are specialized agricultural machines designed for the efficient and precise planting of seedlings or transplants in large-scale farming operations. Mechanical transplanters feature a mechanized planting mechanism that automates the transplanting of seedlings into the soil, enhancing planting efficiency in large agricultural fields. These machines control the depth and spacing of the plantings.

Medium-Duty Truck

A medium-duty truck is a commercial vehicle designed to handle a range of moderate- to heavy-duty applications. Positioned between light-duty trucks and heavy-duty trucks, these vehicles are versatile and find application in various industries. Medium-duty trucks are characterized by their ability to handle moderate to substantial loads, making them suitable for local delivery, construction, and service industries.

Medium-Duty Truck Hydraulic ABS

The medium-duty truck hydraulic ABS uses a booster system that makes sure that the brake assist is out there if the engine stalls or if there is a malfunction within the power steering/brake assist system. Mechanical force is converted into hydraulic pressure by the piston chamber, regulated to satisfy braking system demands by the pressure balance system, and delivered to the brake wheel circuits by the pipes and versatile hoses. For more information on this, see **J2627_200908** [1].

Mid-Size Truck

A mid-size truck is a commercial vehicle that falls between compact and full-size trucks in terms of dimensions, payload capacity, and overall capability. SAE International establishes standards and terminology to ensure clarity and consistency within the automotive industry, particularly concerning these versatile trucks' design, performance, and safety aspects.

Mining Off-Road Vehicle

An off-road mining vehicle is a specialized commercial vehicle designed for operation in challenging and rugged mining environments. Mining off-road vehicles are characterized by robust and durable construction, capable of withstanding extreme conditions commonly encountered in mining operations, including uneven terrain, rocks, and heavy loads. Guidelines may specify standards for the maximum payload capacity, ensuring that mining off-road vehicles can efficiently transport substantial volumes of minerals, ore, or other materials within the mining site. These vehicles are designed with superior off-road capabilities, often featuring AWD or 4WD systems to navigate challenging terrains such as gravel, mud, and steep inclines.

Mini-Truck

A mini-truck is a compact commercial vehicle designed for urban or light-duty applications, combining the versatility of a small cargo vehicle with fuel efficiency and maneuverability.

Mobile Clinic

A mobile workshop is a specialized commercial vehicle equipped with tools, equipment, and facilities for performing on-site maintenance, repairs, or other technical services. A mobile workshop is designed to bring essential tools and services directly to the location where maintenance, repairs, or specialized work is required, eliminating the need to transport vehicles to a fixed workshop.

Mobile Workshop

See **Mobile Clinic**.

Moldboard Plow

A moldboard plow is a specialized agricultural implement designed for soil tillage and preparation. A moldboard plow is a primary tillage tool used to break and turn over the soil in preparation for planting crops. It is particularly effective in burying crop residues and incorporating organic matter into the soil. Guidelines may specify criteria for the design and construction of the moldboard, which is the curved, wedge-shaped blade responsible for turning and burying soil. Standards may cover the attachment mechanisms used to connect the moldboard plow to tractors or other agricultural vehicles. This includes considerations for compatibility, stability, and ease of attachment. Features allow for adjustability, such as the ability to vary the plowing depth, the angle of the moldboard, or the spacing between furrows.

Motorhome

A motorhome is a specialized type of recreational vehicle (RV) that combines living quarters with motorized transportation.

A motorhome features an integrated living space with various amenities, including sleeping quarters, a kitchen, bathroom facilities, and often a dining or lounge area. Unlike towable RVs, motorhomes have their motorized chassis, allowing them to be driven independently without the need for a separate towing vehicle. These vehicles fit classifications for different types of motorhomes, such as Class A, Class B, and Class C, based on size, amenities, and overall design. Standards may cover the powertrain specifications, including engine performance, transmission, and chassis design, ensuring safe and reliable operation in various road conditions. Standards may address the design and safety considerations for electrical and plumbing systems within motorhomes, encompassing features such as generators, water supply, and waste disposal. Guidelines may specify criteria for integrating the living quarters with the motorized chassis to ensure structural integrity, stability, and occupant safety.

Moving Truck

A moving truck is a specialized commercial vehicle designed to transport household or commercial goods during relocations. Moving trucks are equipped with storage compartments, cargo space, and loading ramps or lift gates to facilitate the efficient and safe transportation of belongings from one location to another.

Moving trucks have a large and enclosed cargo space that can vary in size, accommodating a range of items from furniture to boxes. Many moving trucks are equipped with loading ramps or hydraulic lift gates to facilitate the loading and unloading of heavy or bulky items. The cargo area of a moving truck is typically provided with tie-down points to secure items during transit and prevent shifting. The driver's cabin of a moving truck is separated from the cargo area and often includes amenities such as air conditioning, radio, and ergonomic seating. Moving trucks are designed to carry a specific weight capacity, determined by factors such as size and construction.

Mud Pump

A mud pump is specialized equipment designed for the oil and gas industry, particularly in drilling operations. SAE International may provide standards and definitions to ensure clarity and consistency in these commercial pumping units' design, performance, and safety aspects.

Mud pumps are integral components in drilling rigs tasked with circulating drilling fluid (commonly known as mud) into the wellbore during drilling operations. This circulation helps cool the drill bit, carry cuttings to the surface, and maintain wellbore stability. Typically, mud pumps operate on positive displacement principles, ensuring a consistent and controlled flow of drilling fluid under high pressure. Mud pumps are designed to handle high pressures and volumes to meet the demanding requirements of deep drilling operations.

Mud pumps are subjected to harsh conditions, and standards may specify criteria for durability, corrosion resistance, and materials used in their construction. Standards may include specifications for the control systems of mud pumps, ensuring operators can manage pressure, flow rate, and other parameters efficiently.

Mud Tank

A mud tank is a large container or reservoir used in the drilling industry, particularly in oil and gas drilling operations. Mud tanks are designed to hold drilling fluids, also known as drilling mud, which serve various purposes during the drilling process.

Mud tanks can vary in size and capacity, ranging from small to large tanks with thousands of barrels of drilling fluid. Mud tanks are typically constructed from steel and may have internal compartments to separate different drilling fluids. Some mud tanks are equipped with mixing and agitation systems to ensure proper suspension of solids in the drilling fluid. Mud tanks often have a sloped or conical bottom that facilitates the removal of solids and allows for easy cleaning. Mud tanks have openings, valves, and connections for the inflow and outflow of drilling fluids and adding chemicals and additives.

Muffler

A muffler is a component of the exhaust system designed to reduce the noise generated by the exhaust gases expelled from the engine. Mufflers are essential to commercial vehicles to minimize the noise level produced by the engine's exhaust gases, ensuring compliance with noise regulations and providing a more comfortable environment for the driver and surrounding areas.

The primary function of a muffler is to reduce the noise produced by the high-velocity exhaust gases as they exit the engine. Mufflers use techniques, such as internal baffles, chambers, and sound-absorbing materials, to dampen and attenuate the exhaust noise. Mufflers come in various shapes and designs, often cylindrical or rectangular, with internal components to redirect and reduce sound waves. Mufflers are typically constructed from steel or other durable materials that withstand high temperatures and corrosive exhaust gases. Mufflers are installed along the exhaust system, usually near the vehicle's rear, to reduce exhaust noise before exiting the tailpipe.

Multi-Link Suspension

A multi-link suspension is an independent suspension system used in vehicles to enhance ride comfort, handling, and stability. This advanced suspension design uses multiple links or arms to connect the wheel hub to the vehicle chassis, providing better control over wheel movement and allowing each wheel to react independently to road irregularities.

A multi-link suspension system uses multiple control arms or links to connect the wheel hub assembly to the chassis at different attachment points. Each wheel can move independently of the others, enabling better traction and improved ride quality. Multi-link suspensions often offer adjustability in alignment and camber, allowing for fine-tuning handling characteristics. Multi-link suspensions can have complex geometries that optimize wheel movement and control, particularly during cornering and uneven road conditions. Many multi-link suspension systems include anti-roll (sway bars) to reduce body roll during cornering.

REFERENCE

1. Society of Automotive Engineers, "Standard J2627_200908: Braking System," SAE Publishing, Warrendale, 2009.

N

Navigation System

See **GPS** and **Telematics Systems**.

N_2O (Nitrous Oxide)

N_2O, commonly known as laughing gas, is a colorless and odorless gas composed of nitrogen and oxygen. In commercial vehicles, N_2O is often used as a performance-enhancing agent in ICEs. It is introduced into the engine's intake system to increase the available oxygen during combustion, thereby improving engine efficiency and boosting power output.

Off-Road Dump Truck

Off-road dump trucks are specialized heavy-duty vehicles that transport bulk materials in rugged and off-highway terrains, such as earth, rocks, and other construction aggregates. Off-road dump trucks are built with sturdy frames, reinforced chassis, and durable components to withstand the harsh conditions encountered in construction, mining, and quarrying operations. Depending on the design and intended application, these trucks may feature articulated or rigid frames. Articulated trucks provide enhanced maneuverability, while tough trucks offer simplicity and stability.

Off-road dump trucks often have AWD or 4WD systems for optimal traction on challenging and uneven terrains. Standards may include specifications for ground clearance to enable off-road dump trucks to navigate uneven surfaces, obstacles, and rough terrains encountered in mining and construction sites. Off-road dump trucks are typically equipped with powerful engines to ensure sufficient torque and horsepower for heavy loads in demanding conditions. These trucks feature hydraulically operated dumping mechanisms to efficiently unload materials at the job site. SAE guidelines may establish standards for the design and performance of these dumping systems.

Off-Road Vehicle

An off-road vehicle is a specialized vehicle designed and constructed to operate on surfaces other than conventional paved roads. Off-road vehicles are engineered to traverse diverse and challenging terrains, including but not limited to dirt trails, dunes, rocky landscapes, mud, and other uneven surfaces. Many off-road vehicles have 4WD or AWD systems to enhance traction and maneuverability in challenging conditions. Off-road vehicles typically feature elevated ground clearance to navigate obstacles and rough terrains without risking damage to the undercarriage. Off-road vehicles often employ tires with aggressive tread patterns for improved traction on various surfaces. Some off-road vehicles, particularly those designed for extreme conditions, may incorporate rollover protection structures (ROPS) to enhance occupant safety.

Oil Cooler

An oil cooler is a heat exchange device designed to regulate and dissipate excess heat from the engine oil in a commercial vehicle. The oil cooler helps maintain the proper operating temperature of the engine oil, which is essential for optimal engine performance and longevity.

An oil cooler facilitates heat transfer from the engine oil to another medium, often air or coolant, to dissipate excessive heat. Oil coolers can be air-cooled or liquid-cooled depending on the design and the vehicle's configuration. Oil coolers are usually mounted near the engine, either in the engine block or in the radiator area. Oil coolers typically consist of finned tubes or plates that increase the surface area for efficient heat exchange. Some oil cooler systems have a thermostat that regulates oil flow to the cooler based on the oil's temperature.

Oilfield Data Collection Vehicle

An oil field data collection vehicle is a specialized vehicle used in the oil and gas industry to gather, monitor, and transmit various data types from oil and gas field operations. These vehicles have advanced technology and instrumentation to facilitate real-time data collection and analysis for optimizing production processes and decision-making.

Oil field data collection vehicles are equipped with sensors, instruments, and data acquisition systems to capture information related to well performance, reservoir conditions, and production parameters. These vehicles have communication systems, including satellite and wireless networks, to transmit collected data to central control centers or remote analysis facilities. Oil field data collection vehicles may have onboard computing systems capable of processing and analyzing gathered data in real time. Some vehicles have environmental sensors to monitor air quality, temperature, and weather conditions. Safety features are incorporated to ensure the vehicle's safe operation in challenging oil field environments.

Operating Cycle (Engine)

An operating cycle engine is an ICE that undergoes a series of processes to convert fuel into mechanical energy. The operational cycle typically includes intake, compression, power, and exhaust strokes, following the principles of the Otto cycle (commonly used in gasoline engines) or the Diesel cycle (widely used in diesel engines).

Orchard Tractor

Orchard tractors are specialized agricultural vehicles designed for tasks specific to orchard and vineyard operations. Orchard tractors are characterized by compact size, facilitating easy maneuverability through narrow rows of fruit trees or vines in orchards and vineyards. Orchard tractors often include specifications for a low profile to navigate beneath low-hanging branches without causing damage to the crops. These tractors may be equipped with specialized attachments such as pruners, sprayers, or harvesting equipment tailored to the individual needs of orchard and vineyard tasks. Orchard tractors are designed to provide sufficient power for their intended functions while prioritizing fuel efficiency and environmental considerations. Standards for orchard tractors encompass ergonomic designs for operator comfort during extended use and safety features to mitigate risks associated with orchard terrain and equipment operation. Orchard tractors are engineered to navigate diverse terrains found in orchards and vineyards, including uneven surfaces and slopes.

Overdrive

Overdrive refers to a higher gear ratio in a vehicle's transmission that allows the engine to run at a lower speed while maintaining a higher rate of vehicle speed. Overdrive is typically used in the highest and second highest transmission gear, allowing for fuel-efficient and quieter cruising at highway speeds.

Overdrive gears have a higher gear ratio, meaning that the driveshaft revolutions are faster than the engine speed (RPM), which results in a higher vehicle speed at lower engine RPM. By allowing the engine to operate at lower RPMs while maintaining speed, overdrive gears contribute to improved fuel efficiency. Using the engine at lower RPMs in overdrive mode can reduce engine wear and extend life.

Overdrive gears result in quieter engine operation, making a more comfortable driving experience, especially at highway speeds. Vehicles equipped with overdrive can downshift to lower gears when additional power or torque is needed, such as when climbing steep grades.

P

Park Control Valves (Air Brakes)

The spring brake park control valve may be a dash-mounted control valve that permits the operator to use or release the truck or tractor spring parking brakes. The valve model used can vary among different air system circuits. They are generally a push–pull dual-circuit model. The valve button must be pushed in to release the parking brakes by exhausting air from the spring brake chambers. Different models of spring brake chambers are available. If atmospheric pressure is lost within the brake, the buttons will come out. When the vehicle is parked and therefore the dash control button is pulled out, air is exhausted from the hand brake chamber, and therefore, the heavy volute spring forces the brake slack adjuster forward to use the shoes against the drum or restraint against a disc. No air is left within the emergency brake chamber. The hand brake system of a tractor–trailer combination manages hold-off pressure delivered to the spring brake chambers. **FMVSS 121** requires that the shape of every valve is as follows—the color of every valve is an industry standard:

- Red Octagonal Knob: The trailer supply valve. The system spring brake control valve must have already been actuated before this valve may be. It supplies all system air to the trailer(s) and must vent to release the trailer spring brakes.
- Yellow Diamond/Square: The system spring brake (park) control valve. This valve masters the parking brakes on a combination. It is pushed inward to release the parking brakes. When the switch is pulled it is set in park position, it cuts the air supply to the trailer (thus applying its park/emergency brakes) and puts the tractor into park/emergency mode.
- Blue Round: It isolates the tractor park brake function from that of the trailer. This valve is mastered by the spring brake control valve.

Particulates (or Particulate Matter)

The quantity of all solid and liquid particles suspended in air, many of which are hazardous. It is basically soot composed of very small particles of solid/semisolid material. This complex mixture includes both organic and inorganic particles, such as dust, pollen, soot, smoke, and liquid droplets. This is the black soot you sometimes see coming out of an older diesel engine. For more information on this, see **J1145_201109** [1].

Paver

In commercial vehicles, particularly within the construction and road-building industries, pavers refer to specialized machines designed for the placement of asphalt or concrete on road surfaces. There are standards and definitions for these vehicles to ensure consistent terminology and specifications within the automotive industry.

Pavers are equipped with mechanisms for distributing asphalt or concrete onto road surfaces. This process is crucial for achieving even and durable road construction. Pavers incorporate screed systems vital in leveling and compacting the material during the paving process. Pavers are often mounted on crawler tracks or wheels, providing mobility and stability during paving operations to precisely manage the placement of materials. Advanced control systems are integral to pavers, allowing operators to precisely manage the placement of materials. Pavers commonly include heating systems to keep the asphalt or concrete at optimal temperatures for proper placement.

Payload

Payload refers to the total weight of goods, materials, passengers, and any other cargo a commercial vehicle is designed to transport. It is the cumulative mass a vehicle can carry in addition to its curb weight.

In the context of commercial vehicles, payload is a crucial metric that directly influences the operational efficiency and performance of the vehicle. Payload is a dynamic parameter, and understanding its limits is essential for fleet operators, manufacturers, and regulatory bodies. Overloading a vehicle beyond its designated payload capacity can compromise safety, affect handling characteristics, and lead to accelerated wear and tear on various components.

Payload Capacity

Payload capacity refers to the maximum weight a commercial vehicle is designed to carry in addition to its curb weight. This capacity encompasses the total mass of cargo, passengers, and other equipment or materials the vehicle intends to transport.

Payload capacity is a critical specification for commercial vehicles, encompassing a range of applications from trucks and vans to buses. It plays a vital role in determining the efficiency and operational capabilities of the vehicle, impacting factors such as fuel efficiency, braking performance, and overall stability during transit.

SAE standards for payload capacity are established to ensure safety, compliance with regulations, and consistency in rating methodologies across the automotive industry. These standards may consider axle ratings, suspension systems, and structural integrity to accurately determine the vehicle's ability to handle varying loads without compromising safety or performance.

Platooning

Platooning is an innovative concept in commercial vehicles and transportation, particularly in the trucking industry (Figure P.1). It involves a group of vehicles, often trucks or other large commercial vehicles, traveling in a close formation along a highway or roadway. These vehicles are connected through advanced technology, allowing them to communicate and synchronize their movements, often to improve efficiency, safety, and fuel economy [2].

FIGURE P.1 An example of truck platooning.

Courtesey of The Federal Highway Administration.

Platooning typically involves a lead vehicle followed closely by a series of trailing vehicles. The distances between these vehicles are minimized to create a tightly packed formation. Platooning relies on autonomous driving technology, including sensors, cameras, radar, GPS, vehicle-to-vehicle (V2V) communication systems, and advanced control algorithms. These technologies enable the vehicles to operate nearby while maintaining safety. Vehicles within a platoon constantly communicate, sharing information about their speed, acceleration, and braking. This communication allows for synchronized actions among the vehicles. Safety is a paramount consideration in platooning. Advanced safety systems, such as collision avoidance technology and ACC, are employed to prevent accidents and maintain a safe following distance [3].

One of the primary benefits of platooning is improved fuel efficiency. By traveling closely together, the vehicles reduce air resistance, also known as aerodynamic drag. This results in fuel savings for all vehicles in the platoon. In many platooning scenarios, the lead vehicle is operated by a human driver, while the following vehicles can be partially or fully autonomous. This allows drivers in trailing vehicles to disengage from active driving tasks, although they must remain attentive and ready to take control if needed. Platooning can help optimize highway traffic flow by reducing congestion and maintaining a consistent speed. This can lead to smoother traffic and potentially minimize travel times.

Platooning is particularly relevant in long-haul trucking, where fuel costs are significant. It is also explored in freight transportation, logistics, and public transportation. Despite the benefits, platooning faces challenges, including regulatory issues, infrastructure requirements (such as dedicated lanes or roadside equipment), and the need for cybersecurity to protect connected vehicle systems.

Many regions are working on regulations and standards to govern the deployment and operation of platooning technology, ensuring its safety and compatibility with existing road traffic laws. The integration of platooning with regular, non-platooning vehicles poses challenges in providing safe interactions between different types of vehicles on the road.

Police Vehicle

A police vehicle is a specialized commercial vehicle designed and equipped for law enforcement. Law enforcement agencies operate these vehicles and serve various roles, including patrolling, traffic enforcement, emergency response, and transporting personnel or detainees. Police vehicles are equipped with specific features and technologies to support the enforcement of laws and the maintenance of public safety.

Postal Delivery Vehicle

A postal delivery vehicle refers to a specialized vehicle used by postal services to transport and deliver mail, packages, and parcels to residential and business addresses. These vehicles are designed to facilitate the efficient and timely distribution of postal items within specific routes or areas.

Postal delivery vehicles have dedicated cargo compartments designed to accommodate various sizes of mail items, packages, and parcels. These vehicles often have sliding doors, roll-up doors, or hinged compartments that provide easy access to mail for delivery and collection. The interior of postal delivery vehicles may feature shelves, bins, dividers, and sorting compartments to help postal workers organize mail items for efficient delivery. Many postal delivery vehicles include security measures, such as locking compartments, to protect the safety and integrity of mail items during transportation. Postal delivery vehicles are designed with considerations for visibility and maneuverability as they navigate urban and suburban areas during mail distribution (Figure P.2).

FIGURE P.2 Mail delivery requires a balance of cargo-carrying capacity and ease of movement.

Johnny Habell/Shutterstock.com.

Postal Truck

A postal truck is a specialized vehicle for postal services to deliver mail and packages to residential and business addresses. Postal trucks are designed to transport mail items while accommodating the unique requirements of postal operations efficiently and safely.

Postal trucks are equipped with dedicated cargo compartments or compartments adapted for carrying various sizes of mail items, parcels, and packages. These trucks often have sliding doors, roll-up doors, or hinged compartments to facilitate easy access to mail for delivery and collection. Interior configurations include shelves, bins, and dividers to help postal workers organize mail items for efficient sorting and delivery.

Many postal trucks have security measures, such as locking compartments, to ensure the safety of mail items during transport and delivery. Postal trucks may have large windows and mirrors to give drivers optimal visibility while maneuvering in urban and suburban areas.

Potato Harvester

A potato harvester is a specialized agricultural machine designed for the efficient and mechanized harvesting of potatoes. Typically attached to or towed by commercial vehicles, such as tractors, these harvesters play a crucial role in modern potato farming by automating the process of digging, lifting, and separating potatoes from the soil, vines, and other debris (Figure P.3).

FIGURE P.3 A potato harvester harvesting and loading potatoes into the waiting vehicle.

Maksim Safaniuk/Shutterstock.com.

Potato harvesters have digging blades or digger shares that penetrate the soil to lift the potato plants from the ground. The lifted plants are then conveyed to the surface. Once the potato plants are raised, the harvester has mechanisms for separating the potatoes from the soil, stones, and debris. This often involves a series of shaking screens, conveyors, and rollers. As potatoes are separated from the soil, they are collected and conveyed to a holding area or a trailer attached to the harvester. This area typically includes storage bins or hoppers. Potato harvesters have adjustable digging depths to accommodate variations in soil conditions and to ensure that all potatoes are harvested without damaging them.

Many harvesters are designed with gentle handling mechanisms, including rubberized or padded conveyors and rollers, to minimize damage to the potatoes during harvesting. Some potato harvesters have size grading options that sort harvested potatoes into different size categories, making it easier for farmers to market their produce. Potato plants often have vines or haulms attached to them. Potato harvesters include mechanisms to separate the potatoes from these vines, leaving only the tubers for collection.

Potato harvesters significantly reduce the labor required for manual potato harvesting, making large-scale commercial potato farming more efficient and cost-effective. Potato harvesters can be adjusted to accommodate row widths, allowing flexibility for other farming practices and field layouts. Potato harvesters are often attached to tractors or commercial vehicles using a three-point hitch or a specialized mounting system. Some potato harvesters are equipped with topping systems that remove the upper portion of the potato plants (the foliage) before lifting the tubers. This can help reduce the risk of disease and improve storage quality. Once the holding area is full, the potatoes are dumped into storage bins or trailers for transport to storage facilities or processing plants.

Power-Only Trailer

A power-only trailer is designed to be hauled by a tractor unit or truck without carrying its load. Unlike traditional trailers that carry cargo directly, a power-only trailer transports other trailers or containers that require movement from one location to another. Power-only trailers are often used in logistics and transportation scenarios where the focus is on moving the trailer rather than its contents.

A power-only trailer is designed to transport other trailers, containers, or similar units. It does not have its own cargo capacity. Power-only trailers have a hitch or coupling mechanism that allows them to be connected to a tractor unit or truck for towing. Power-only trailers are commonly used in intermodal transportation to move containers between different modes of transport, such as from a ship to a rail yard or truck. Power-only trailers are connected to the towing vehicle using a fifth-wheel coupling or similar mechanism. Power-only trailers are versatile and can transport many trailers, including dry vans, flatbeds, and refrigerated trailers.

Powertrain

See **Driveline**.

Precision Agriculture

A farming approach that utilizes GPS technology to optimize planting, irrigation, and other farming operations for increased efficiency and yield.

Precision Planter

Precision planters are advanced agricultural implements commonly used with commercial vehicles, particularly tractors, to accurately and efficiently plant crops in rows with specific spacing. These planting machines are a critical component of modern agriculture, as they help optimize seed placement, spacing, and depth, ultimately improving crop yields.

Precision planters are designed to place seeds in the soil at a consistent and precise depth, ensuring optimal germination and crop growth conditions. These planters offer adjustable row spacing, allowing farmers to plant crops at specific intervals based on crop type and farming practices. Standard row spacings include 30 in. for corn and narrower spacings for crops like soybeans. Precision planters use seed metering mechanisms to control the rate at which seeds are released and planted. This ensures that seeds are evenly spaced and cultivated at the desired population per acre. Operators can adjust the planting depth to match the requirements of different crops and soil conditions. Proper depth control is essential for uniform seed germination and root development. Some precision planters are equipped with technology that enables variable rate planting. The planter can adjust seed spacing and planting rates based on field conditions and soil characteristics, optimizing seed usage and crop yield potential. Many precision planters can simultaneously apply fertilizers or soil conditioners at the same time seeds are planted. This precise placement enhances nutrient availability for young plants.

GPS technology is often integrated into precision planters, allowing for accurate row placement and field mapping. This technology can also facilitate straight and uniform rows, reducing overlap and waste. Modern precision planters have seed monitoring systems that track seed delivery, allowing operators to identify and address planting issues or blockages quickly. Precision planters can be integrated with other agricultural technologies, such as GPS-based guidance systems and data management software. This integration enables more efficient and data-driven farming practices. Some precision planters use advanced singulation mechanisms to separate and place individual seeds at precise intervals, reducing seed waste and improving crop stand uniformity. Electric drive systems are becoming increasingly common in precision planters, offering greater control over seed metering and spacing and improved planting accuracy. Routine maintenance of precision planters, including cleaning, calibration, and regular lubrication, is essential to ensure reliable performance.

Precision Planting Equipment

Farm machinery is equipped with GPS technology and software to enable precise planting operations.

Prescription Planting

GPS data and soil information create customized planting prescriptions that optimize seed placement and spacing.

PTO (Power Take-Off)

A PTO is a mechanical device that transfers power from a vehicle's engine to auxiliary equipment or attachments. These attachments can be in front of the vehicle or at the back. PTOs are commonly found in various types of commercial vehicles, such as trucks, buses, and utility vehicles, and they enable the use of additional equipment that requires engine power, such as hydraulic pumps, generators, compressors, and winches.

A PTO connects directly to the vehicle's transmission, typically using a drive shaft, gears, and clutches to tap into the engine's power. PTOs transfer power to various auxiliary equipment, including hydraulic pumps for operating cranes, compressors for air brake systems, generators for power supply, and winches for recovery operations. Many vehicles equipped with PTOs have a control mechanism that engages and disengages the PTO as needed. There are various types of PTOs, including rear-mounted PTOs, side-mounted PTOs, and split-shaft PTOs, each suited for specific applications.

Pusher Axle

The pusher axle refers to an auxiliary axle added, typically a truck or a truck–trailer combination. The purpose of a pusher axle is to provide additional load-bearing capacity and distribute the weight of the cargo more evenly, improving the vehicle's weight distribution and compliance with legal weight limits. The pusher axle is very identical to the tag axle with the exception of the pusher axle located in front of the drive axle(s).

A pusher axle is an extra axle that is not part of the vehicle's primary configuration but is added to increase the overall number of axles. Some pusher axles are designed to be retractable or liftable, meaning they can be raised when not needed to reduce tire wear and improve fuel efficiency. Many pusher axles have air suspension systems for better ride comfort and adjustability. Pusher axles are designed to support a certain amount of weight, contributing to the overall load-carrying capacity of the vehicle. Pusher axles help commercial vehicles meet legal weight limits imposed by regulations, preventing overloading.

REFERENCES

1. Society of Automotive Engineers, "J1145_201109: Emissions Terminology and Nomenclature," SAE Publishing, Warrendale, 2011.

2. Ortega, J. and Salari, K., "Experimental Investigation of the Aerodynamic Benefits of Truck Platooning: Two- and Four Vehicle Platoons," *SAE Int. J. Adv. & Curr. Prac. in Mobility* 3, no. 6 (2021): 2782-2801, doi:https://doi.org/10.4271/2021-01-0942.

3. McAuliffe, B., Raeesi, A., Lammert, M., Smith, P. et al., "Impact of Mixed Traffic on the Energy Savings of a Truck Platoon," *SAE Int. J. Adv. & Prac. in Mobility* 2, no. 3 (2020): 1472-1496, doi:https://doi.org/10.4271/2020-01-0679.

R

Rear Axle

The back axle of a car or truck. The rear axle transfers power from the engine to the wheels, and carries the weight of the vehicle.

Recirculating Ball Steering

Recirculating ball steering is a type of steering system commonly used in more extensive and heavier commercial vehicles, such as trucks, buses, and recreational vehicles. This steering system is designed to provide precise control and efficient steering for vehicles that require more significant effort due to their weight and size. Recirculating ball steering systems use a combination of gears and balls to transmit steering input from the driver to the wheels.

The core component of the recirculating ball steering system is the steering gearbox, which contains gears and a recirculating ball mechanism. Ball bearings reduce friction and enable smooth movement of the gears within the steering gearbox. The sector shaft is connected to the steering wheel and moves the recirculating ball mechanism, translating driver input into steering motion. The pitman arm is connected to the sector shaft and transfers the motion to the steering linkage, eventually moving the wheels. The recirculating ball mechanism consists of a ball nut and a worm gear. As the sector shaft rotates, it impacts the ball nut, causing the balls to recirculate and translate the rotational motion into linear motion. Recirculating ball steering systems allow adjustment to achieve the desired level of steering play and responsiveness.

Reefer Truck

See **Refrigerated Truck**.

Refrigerated Trailer

A refrigerated trailer is a specialized type equipped with refrigeration or cooling systems designed to maintain a controlled temperature environment for transporting perishable goods and temperature-sensitive cargo.

Refrigerated trailers are equipped with refrigeration units that allow operators to control and maintain specific temperature ranges within the trailer's cargo area. These trailers have well-insulated walls, floors, and ceilings to minimize temperature fluctuations and ensure efficient cooling. Refrigerated trailers feature self-contained refrigeration units that use compressors, evaporators, and condensers to cool and circulate air within the cargo space. Many modern refrigerated trailers have temperature monitoring and recording systems that provide real-time data and ensure compliance with temperature requirements. Some refrigerated trailers have loading ramps or lift gates for efficient loading and unloading of cargo.

Refrigerated Truck

A refrigerated or reefer truck is a specialized commercial vehicle designed to transport perishable goods and temperature-sensitive cargo at controlled and consistent temperatures. Refrigerated trucks are equipped with refrigeration systems that maintain a specific temperature range to preserve the quality and freshness of the transported goods throughout the journey.

Refrigerated trucks have insulated walls, floors, and ceilings to provide thermal insulation, helping to maintain the desired internal temperature. These trucks are equipped with refrigeration units that use a compressor, evaporator, condenser, and refrigerant to cool and actively regulate the internal temperature. The refrigeration system can be adjusted to maintain specific temperature ranges depending on the transported cargo. Air circulation systems ensure even cold air distribution within the cargo area to prevent temperature fluctuations and maintain consistent conditions.

Many refrigerated trucks have temperature monitoring systems that allow operators to monitor and adjust the temperature remotely. Some refrigerated trucks have built-in loading ramps to facilitate the efficient loading and unloading of goods.

Regeneration

Regeneration is used in diesel engine vehicles with DPFs. A DPF is a component designed to capture and reduce the emission of PM (soot) from the exhaust gases of a diesel engine. Over time, the DPF can become clogged with trapped soot, affecting engine performance and emissions. Regeneration means burning off or removing the accumulated soot from the DPF to restore its efficiency and functionality.

DPF is a device installed in the exhaust system of a diesel engine that traps and stores PM. As the engine operates, soot particles accumulate within the DPF, gradually reducing its ability to filter exhaust gases effectively. Regeneration involves raising the temperature within the DPF to a level where the collected soot is burned off and converted into ash. This process typically occurs during regular engine operation. In some vehicles, active regeneration is initiated by the ECU to artificially increase exhaust temperature and promote soot combustion. Passive regeneration occurs naturally due to normal driving conditions, where exhaust temperatures are naturally elevated, promoting soot burn-off.

Remote Vehicle Diagnostics

These systems allow real-time vehicle health monitoring, detecting faults and potential issues before they become significant problems, and reducing downtime and maintenance costs.

RFID (Radio Frequency Identification)

RFID technology uses radio waves to identify and track objects or assets, including commercial vehicles, by electronically storing and transmitting information. RFID systems consist of tags, readers, and software infrastructure that enable wireless data communication between tags and readers. RFID technology has various applications in managing and tracking commercial vehicles, cargo, and logistics operations.

RFID tags are small devices with an electronic chip and antenna. They can be attached to commercial vehicles, containers, packages, or assets. RFID readers emit radio waves to communicate with nearby RFID tags and retrieve information from them. RFID systems operate using different frequency bands, such as low frequency (LF), high frequency (HF), and ultra-HF (UHF), each with its range and characteristics. The read range of RFID systems varies based on the frequency and technology used, ranging from a few centimeters to several meters. RFID tags can be passive (powered by the reader's energy) or active (equipped with their power source), affecting their range and capabilities. RFID tags can store various types of information, such as identification numbers, product details, maintenance records, and more.

RGN (Removable Gooseneck) Trailer

An RGN trailer refers to a specialized type of flatbed trailer designed for transporting oversized and heavy loads, particularly equipment and machinery. The defining feature of an RGN trailer is its detachable front section, known as the gooseneck. This section can be hydraulically raised and lowered to create a ramp for loading and unloading equipment. RGN trailers typically have a low deck height near the trailer's rear, enabling taller equipment to be loaded without exceeding height restrictions. RGN trailers often have multiple axles to distribute heavy loads' weight and provide better transportation stability.

RGN trailers transport a wide range of oversized cargo, including construction equipment, large machinery, and industrial components. Some RGN trailers have extendable deck sections that can be adjusted to accommodate longer items. RGN trailers come in various lengths and widths to cater to different types of loads. RGN trailers have safety features such as tie-down points, loading ramps, and lighting systems for safe and secure transportation.

Ride-Sharing Vehicles (Uber and Lyft)

A ride-sharing vehicle is a commercial vehicle used in the ride-sharing industry where passengers can request transportation services through a digital platform or mobile application. Companies like Uber and Lyft operate ride-sharing services, connecting drivers with passengers seeking transportation from one location to another. Ride-sharing vehicles are typically owned by independent drivers who partner with ride-sharing platforms to offer on-demand transportation services.

Risk Management

Risk management systems help identify and mitigate risks associated with vehicle operations, safety, and compliance, reducing potential liabilities.

Roof Rack

A roof rack refers to a framework or structure mounted on the roof of a vehicle, such as a car, Sports Utility Vehicle (SUV), van, or truck. Roof racks are designed to carry and transport additional cargo, equipment, or items that cannot fit inside the vehicle's interior. They provide a practical solution for increasing a vehicle's carrying capacity, especially for larger or bulkier items.

A roof rack consists of crossbars, supports, and attachments that create a platform on the vehicle's roof. Roof racks are securely mounted to the vehicle's roof using various attachment methods, such as clamps, brackets, or integrated mounting points. Many roof racks allow crossbars to be adjusted to accommodate different cargo sizes and shapes.

Roof racks are available in various configurations, including crossbars with integrated accessories like bike mounts, cargo boxes, and kayak carriers. Roof racks are designed to handle specific weight capacities, and adhering to the manufacturer's recommendations for safe loading is essential. Many roof rack accessories come with locks or security features to prevent the theft of items.

ROPS (Rollover Protective Structure)

ROPS is a safety feature designed to protect operators of certain commercial vehicles, particularly those used in agriculture and construction, in the event of a rollover accident (Figure R.1). ROPS consists of a reinforced frame or protective structure installed on the vehicle to minimize the risk of injury to the operator by providing a protective zone in the event of a rollover or overturn [1].

ROPS structures must comply with established certification standards and regulations, ensuring they meet safety requirements and can effectively protect the operator during a rollover. ROPS is often designed to work with other safety features, such as seat belts and operator restraint systems, to provide comprehensive protection during a rollover event.

For more information, see **J1119** [3].

FIGURE R.1 Examples of ROPs for a tractor [2].

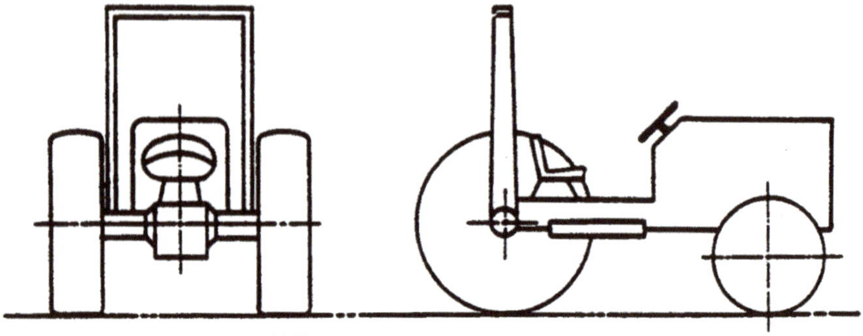

(a) Tractor with typical two-post ROPs

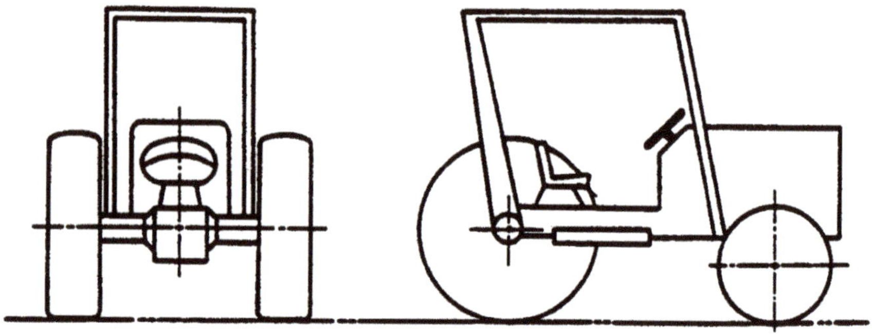

(b) Tractor with typical four-post ROPS

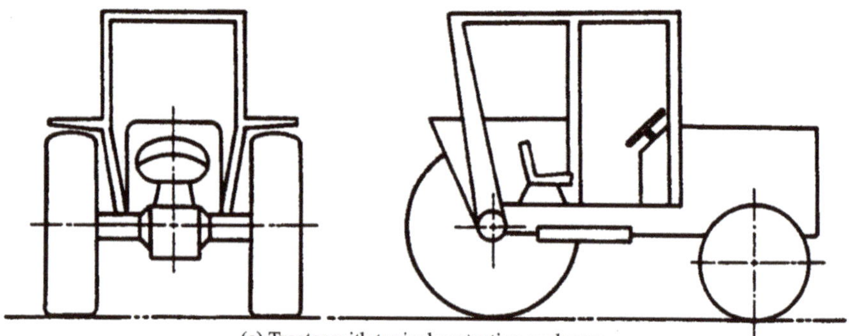

(c) Tractor with typical protective enclosure

Rotary Mower

Rotary mowers, often called rotary cutters or brush hogs, are essential in commercial vehicles, particularly those used in agriculture, landscaping, construction, and roadside maintenance (Figure R.2). These mowers are designed to cut efficiently and clear grass, weeds, brush, and other vegetation in various environments and applications.

FIGURE R.2 There are many variations on mowers, not just rotary; these fit behind the propelling equipment [4].

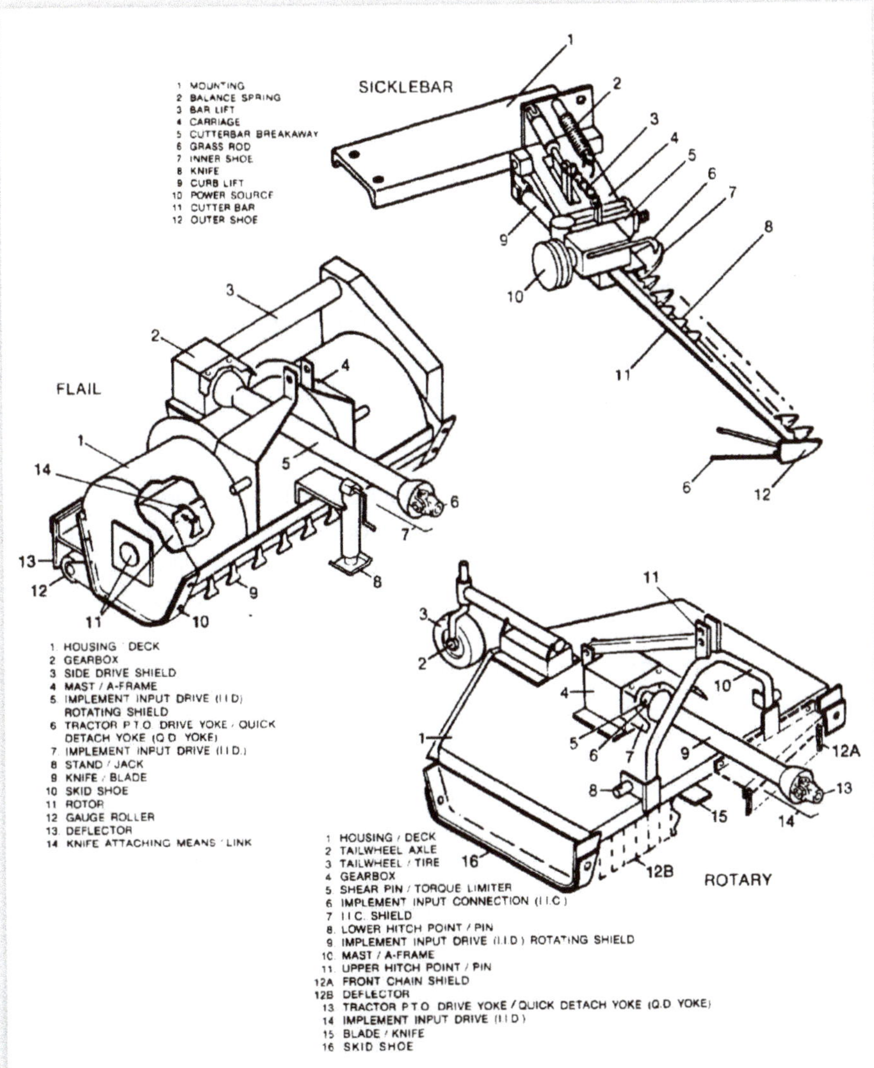

Rotary mowers feature a horizontally rotating blade or blades that operate parallel to the ground. These blades use a slicing or chopping action to cut vegetation. Rotary mowers can be powered by various sources, including PTO from the vehicle's engine, hydraulic systems, or independent gasoline or diesel engines mounted on the mower. Using a three-point hitch or other suitable mounting systems, rotary mowers are typically attached to commercial vehicles, such as tractors, skid steer loaders, excavators, or trucks. These mowers come with different blades, including straight blades for cutting grass and light vegetation, heavy-duty blades for thick brush and small trees, and flail-type blades for mulching and finer cutting. Rotary mowers are available in various cutting widths, allowing users to select the appropriate size for the specific task and vehicle. Most rotary mowers offer adjustable cutting heights, enabling operators to control the desired height of the cut vegetation. Rotary mowers can operate on various terrains, including flat fields, slopes, ditches, and rough or uneven surfaces.

Rotary Tiller

Rotary tillers, rototillers, or rotary cultivators are essential agricultural implements used in commercial vehicles, particularly tractors. These powerful machines are designed to prepare and cultivate soil for planting by breaking up compacted soil, incorporating organic matter, and creating a finely textured seedbed.

Rotary tillers are equipped with a set of rotating tines or blades that are mounted on a horizontal shaft. These tines dig into the soil and break it up as the machine is pulled or driven forward. Rotary tillers are powered by the vehicle's PTO, a hydraulic system, or an independent engine mounted on the tiller. The power source drives the tines to perform the soil cultivation. Rotary tillers are typically attached to the rear of a tractor or other compatible commercial vehicles using a three-point hitch or specialized mounting system. Operators can adjust the tilling depth by changing the angle and positioning of the tiller. This allows for precise soil preparation, whether for shallow cultivation or deep plowing.

Rotary tillers come in various widths, ranging from a few feet to several feet wide. The choice of width depends on the tractor's size and the agricultural operation's specific needs. Some rotary tillers are designed with counter-rotating tines, where one set of tines rotates in the opposite direction of the other set. This counter-rotation enhances the tiller's ability to break up compacted soil and improve soil structure. Rotary tillers break up soil and mix organic matter, compost, or other soil amendments into the earth, enhancing soil fertility and structure. Rotary tillers create a finely tilled seedbed ideal for planting or transplanting seedlings. The smooth and level surface promotes uniform seed germination and root growth.

By thoroughly disrupting the soil, rotary tillers can help control weeds by burying weed seeds and disturbing established weed root systems. In agriculture, rotary tillers can incorporate crop residues into the soil after harvest, improving soil health and organic matter content. Routine maintenance, such as lubricating the tiller's moving parts, inspecting belts and blades, and checking the power source, is essential to ensure efficient and trouble-free operation. Operators of rotary tillers should be trained in safe operation practices. Safety shields and guards are typically included to protect against contact with the rotating tines.

Rough Terrain Crane

A rough terrain crane is a specialized mobile crane designed for lifting and moving heavy loads in challenging and uneven terrains. These cranes are used in construction, industrial, and off-road settings where traditional mobile cranes may struggle due to rough, unstable, or uneven ground conditions.

Rough terrain cranes have sturdy, off-road capabilities, including more extensive and rugged tires, powerful engines, and durable suspension systems that allow them to navigate rough, unpaved, and uneven terrains. These cranes feature a hydraulic telescoping boom extending to various lengths to reach and lift loads at different heights and distances. Rough terrain cranes have counterweights to stabilize and prevent tipping while lifting heavy loads.

The crane's operator cabin is typically on top of the rotating upper structure, providing good visibility of the load and surroundings. Many rough terrain cranes have four-wheel steering, allowing them to navigate tight spaces and maneuver on challenging terrains. Extendable outriggers stabilize the crane during lifting operations by providing additional support and preventing tilting. Rough terrain cranes have load charts that offer guidelines for safe lifting based on factors such as boom length, angle, and load weight.

Round Baler

A commercial round baler is an agricultural machine that creates round bales of crop materials, typically forage crops like hay, straw, or silage. It collects loose crop material from the field, compresses it into a cylindrical shape, and wraps it with twine, netting, or other binding materials to create a tightly packed and compact bale.

Round balers have a pickup system equipped with rotating tines or belts that gather the crop material from the ground and feed it into the baling chamber. This is the baler part, where the crop material is compressed and formed into a round bale. It usually consists of rolls or belts that compress the material as it passes through. After the material is compressed, it secures the bale by wrapping it with twine, netting, or other binding material. Some round balers also incorporate automatic or semi-automatic binding systems. Round balers are typically towed by a tractor, equipped with a hitch or connection point to pull them behind the tractor. Many round balers have hydraulic systems that control various functions, such as bale density and size adjustment, as well as opening and closing the rear gate of the baler to release the finished bale. Modern round balers may feature a control panel with settings for bale size, density, and other parameters, which the operator can adjust.

Commercial round balers are essential agriculture tools, particularly in livestock farming, where forage crops are baled for winter feed or storage. They enable efficient and consistent bale production, making handling and transporting bulky crop materials easier. These machines come in various sizes and configurations to accommodate different types of crops and field conditions.

Route Optimization Software

Route optimization software uses algorithms to determine the most efficient routes for commercial vehicles, minimizing fuel consumption, travel time, and operational costs.

Row Crop Tractor

A commercial row crop tractor is a specialized agricultural vehicle designed explicitly for cultivating, planting, maintaining, and harvesting row crops such as corn, soybeans, cotton, and various vegetables. These tractors are characterized by their narrow wheelbase and adjustable wheel spacing, which allows them to navigate between rows of crops with minimal damage to plants.

Row crop tractors have a relatively narrow wheelbase to fit between rows of crops, typically ranging from 60 to 80 in. (1.5 to 2 m) to match the row spacing of the specific crop. Many row crop tractors have axles that can be adjusted in width to fit the row spacing of different crops, allowing for flexibility in farming operations. Row crop tractors often have high ground clearance to accommodate taller crops and prevent plant damage. Most row crop tractors feature front-wheel steering, providing precise control and maneuverability in narrow spaces. Hydraulic systems are commonly integrated into row crop tractors to power various implements and attachments for planting, cultivating, and harvesting row crops. Row crop tractors are equipped with a PTO that provides power to various agricultural implements, such as planters, cultivators, and harvesters. Some row crop tractors have enclosed cabs with climate control for operator comfort and modern control systems for ease of use.

Row crop tractors are essential for farmers who grow crops in rows, as they provide the precision and flexibility needed to manage and maintain these crops efficiently throughout the growing season. These tractors can be equipped with various attachments and implements to perform tasks such as planting seeds, cultivating soil, applying fertilizers and pesticides, and harvesting crops.

Row Cultivator

A commercial row cultivator is an agricultural machine designed to cultivate row crops, such as corn, soybeans, and vegetables. This vehicle typically consists of a frame or chassis equipped with cultivating tools, such as sweeps, shovels, or rotary hoes, that are positioned to work between the rows of crops.

The frame of the cultivator is often mounted on wheels, allowing it to be pulled behind a tractor or other suitable power source. Row cultivators have one or more rows of cultivating tools, which can vary in type and design. These tools are positioned to interact with the soil between crop rows to remove weeds, break up soil crust, and create a weed-free zone around the crop. Many row cultivators have an adjustable width feature, allowing farmers to match the spacing of the cultivating tools with the row spacing of their crops. Depth control mechanisms or wheels enable operators to set the depth at which the cultivating tools work in the soil. Row cultivators are typically pulled by a tractor or another vehicle, which provides the necessary power to drive the cultivator and its cultivating tools.

Row cultivators are crucial in weed management and soil preparation in row-cropped fields. By maintaining a weed-free area around the crop rows, row cultivators help reduce competition for water and nutrients, ultimately promoting better crop growth and higher yields. Additionally, they can break up compacted soil and improve soil aeration, which benefits crop root development.

Running Board

A running board refers to a flat, often step-like, platform mounted to the side of a vehicle, typically just below the doors. Running boards provide a convenient step for passengers entering or exiting the vehicle, especially for larger or taller vehicles. They also serve as a functional and sometimes decorative accessory for commercial vehicles.

Running boards are designed to assist passengers in stepping up or down from the vehicle's cabin. This is particularly useful for taller vehicles or those with higher ground clearance. Some running boards are designed with aesthetic considerations, enhancing the vehicle's appearance and style. Running boards can offer limited protection to the vehicle's undercarriage by shielding it from road debris and splashes.

Running boards can be customized to match the vehicle's design and may incorporate aluminum, stainless steel, plastic, or rubber. Running boards come in various designs, including full-length, cab-length, retractable, and illuminated options. Quality running boards are constructed to withstand various weather conditions and regular use without deteriorating. Full-length running boards extend along the entire length of the vehicle, providing step access to all doors. Cab-length running boards run along the length of the cabin area, giving access to the front doors.

Some models feature running boards that automatically extend (retractable running boards) when a door is opened and retract when the door is closed. These running boards incorporate lighting (illuminated running boards) elements for enhanced visibility during entry and exit, especially in low-light conditions.

REFERENCES

1. Society of Automotive Engineers, "J2194 Roll-Over Protective Structures (ROPS) for Wheeled Agricultural Tractors," SAE Publishing, Warrendale, 2016.
2. Society of Automotive Engineers, "J1194 Rollover Protective Structures (ROPS) for Wheeled Agricultural Tractors," SAE Publishing, Warrendale, 2016.
3. Society of Automotive Engineers, "J1119 Steel Products for Rollover Protective Structures (ROPS) and Falling Object Protective Structures (FOPS)," SAE Publishing, Warrendale, 2013.
4. Society of Automotive Engineers, "J990 Nomenclature - Industrial and Agricultural Mowers," SAE Publishing, Warrendale, 2020.

S

SAE

SAE is a global professional association that serves as a platform for professionals in the automotive and mobility industries. SAE is dedicated to advancing the knowledge, standards, and practices of automotive engineering and technology. Established in 1905, SAE International has become a leading organization that collaborates with industry experts, engineers, researchers, and practitioners to address challenges, develop standards, and promote innovation within the automotive and commercial vehicle sectors.

SAE is a global organization with members and activities spanning the automotive and mobility sectors worldwide. It provides a platform for networking and collaboration among professionals in the field. SAE plays a crucial role in developing and maintaining technical standards and recommended practices for various aspects of automotive engineering, including commercial vehicles. These standards contribute to uniformity, safety, and innovation within the industry.

SAE publishes technical papers, journals, books, and magazines that cover a wide range of topics in automotive engineering, technology, and innovation. These publications are valuable resources for professionals seeking the latest research and insights. SAE collaborates with industry stakeholders, including manufacturers, suppliers, government agencies, and academia, to address common challenges, promote technological advancements, and ensure the industry's competitiveness. SAE advocates for innovation and technical advances in automotive engineering. It serves as a voice for the industry, contributing to developing regulations, policies, and standards that impact the automotive and commercial vehicle sectors.

Safety Equipment

Safety equipment refers to a wide range of specialized devices, systems, and features designed to enhance the safety of both the vehicle occupants and others on the road. Commercial vehicles have various safety equipment to prevent accidents, mitigate their impact, and protect drivers, passengers, pedestrians, and other road users.

Commercial vehicles have airbag systems that deploy upon impact to cushion and protect the occupants during a collision. ABS prevents wheel lockup during braking, allowing the driver to maintain steering control and avoid skidding. ESC helps drivers maintain control during sudden maneuvers or slippery conditions by automatically adjusting braking and engine power. Traction control systems reduce wheel spin during acceleration, improving stability and preventing skidding. Collision Avoidance Systems: Sensors and cameras are used to alert drivers to potential collisions, and sometimes they are intervened to avert or mitigate collisions. LDW alerts drivers if the vehicle drifts out of its lane without using a turn signal. BSD is a system that alerts drivers to vehicles or obstacles in their blind spots, reducing the risk of lane-changing collisions (Figure S.1).

FIGURE S.1 Safety alert symbol that can be found in vehicle instrumentation [1].

AEB systems automatically apply brakes to prevent or mitigate collisions if the driver does not respond to an impending collision. ACC maintains a safe following distance from the vehicle ahead by automatically adjusting the vehicle's speed. Backup cameras provide a clear view of the area behind the vehicle when reversing, reducing the risk of collisions with obstacles.

Pedestrian Detection: Some vehicles are equipped with systems that detect pedestrians and provide warnings or automatic braking if a collision is imminent. The roll stability control system helps prevent rollovers by applying individual wheel brakes and adjusting engine power to maintain vehicle stability. Commercial vehicles may have emergency exit windows, doors, and roof hatches for quick evacuation.

Safety Glass

Safety glass refers to a type of glass used to construct vehicle windows, windshields, and other transparent panels. Safety glass is designed to minimize the risk of injuries in the event of breakage by preventing the glass from shattering into sharp, dangerous fragments. It is a crucial safety feature in vehicles, helping to protect occupants and pedestrians from potential harm.

One common type of safety glass used in windshields is laminated glass. It consists of two layers of glass bonded together with a layer of plastic (usually polyvinyl butyral or PVB) in between. In the event of impact, the glass may crack, but the plastic layer holds the pieces together, preventing them from scattering. Tempered glass is another type of safety glass used in side and rear windows. It is treated with heat and rapid cooling to create internal stresses, making it more breaking-resistant. When tempered glass breaks, it shatters into small, relatively harmless pieces rather than sharp shards.

Safety Rating

Safety rating refers to evaluating or assessing a commercial vehicle's safety performance and compliance, typically conducted by regulatory authorities, organizations, or agencies responsible for transportation safety. A safety rating aims to inform the public, fleet operators, and other stakeholders about the safety practices and standards of a commercial vehicle or carrier.

Safety ratings may consider the overall condition of the commercial vehicle, including mechanical components, structural integrity, and equipment functionality. Some safety ratings incorporate data on driver behavior, such as compliance with traffic laws, HOS regulations, and adherence to safe driving practices. Past accident records, violations, and crash statistics may influence a vehicle's safety rating. The presence and proper functioning of safety technologies, such as anti-lock brakes, stability control, and ADAS, can impact the rating. Regular maintenance, inspection schedules, and adherence to manufacturer guidelines can positively affect a vehicle's safety rating. For carriers transporting cargo, properly securing loads to prevent shifting and falling is critical to safety ratings.

Safety ratings often include different categories or levels to provide a comprehensive assessment. Excellent or superior indicates high safety performance and compliance, demonstrating a strong commitment to safety practices. Satisfactory or acceptable suggests a reasonable level of safety performance with room for improvement in certain areas. Conditional implies that the vehicle or carrier meets minimum safety requirements but has some deficiencies that must be addressed. Unsatisfactory indicates serious safety violations or non-compliance, potentially leading to restricted operation or penalties.

Safety ratings inform the public and shippers about the safety practices of carriers, influencing their decisions when choosing transportation services. Regulatory authorities often use safety ratings to determine a carrier's operating permit and license eligibility. Carriers with higher safety ratings may benefit from more favorable insurance rates. Maintaining a positive safety rating can enhance a carrier's reputation and credibility within the transportation industry. Law enforcement and regulatory agencies may use safety ratings to determine inspection priorities during roadside checks.

Sand Hauler (for Transporting Proppants)

Sand haulers refer to specialized commercial vehicles or trucking companies that transport sand, particularly frac sand, used in the oil and gas industry for hydraulic fracturing operations. Sand haulers are crucial in delivering the necessary proppant materials to well sites, ensuring smooth and efficient fracking processes.

Sand haulers often use specialized trailers, such as pneumatic or hopper trailers, designed to transport bulk quantities of sand safely and efficiently. Sand haulers transport large amounts of frac sand, typically stored in the trailer's hopper compartments, from sand mines or processing facilities to well sites. Sand haulers may be equipped with loading and unloading equipment, such as pneumatic systems or conveyor belts, to facilitate the efficient transfer of sand. Due to the potential hazards of handling bulk materials, sand haulers prioritize safety measures for loading, transportation, and unloading.

Saturation Mapping

A saturation map uses imaging data represented in levels of color intensity based upon levels of an attribute found within an image area. It helps to identify the areas that contain the highest and lowest concentration of that attribute. This will show up as a color gradient within the image. Saturation mapping is a technique that uses GPS data to identify, for example, areas of fields where excessive planting or over-fertilization may occur.

School Bus

A school bus is a specialized type of commercial vehicle designed primarily for the transportation of students to and from educational institutions, such as schools and colleges. School buses are distinct in their design, safety features, and regulatory requirements to ensure the safety and well-being of the passengers, especially children.

School buses are commonly painted bright yellow, which is easily recognizable and aids in increasing visibility and safety on the road (Figure S.2). School buses come in various sizes and configurations, typically seating around 10 to 90 passengers, depending on the model. School buses have various safety features, including high-backed, padded seats, reinforced construction, emergency exits, and reflective markings.

School buses must have multiple emergency exits, including rear and side doors, roof hatches, and windows designed for quick escape in an emergency. School buses are equipped with flashing lights on the front and rear, which indicate stops and alert other drivers when children are boarding or disembarking. A stop arm is a retractable arm with flashing lights attached to the side of the bus. It is extended when the bus stops to pick up or drop off students, indicating that other vehicles should also cease. Some school buses are equipped with a crossing gate at the front, designed to prevent students from crossing in front of the bus where they may not be visible to the driver.

FIGURE S.2 School buses can be built off heavy vehicle chassis and carry children to school and events.

SCR (Selective Catalytic Reduction)

SCR is an advanced emission control technology used in diesel engines to reduce NOx emissions. It involves injecting a precisely dosed liquid urea solution, DEF, or AdBlue, into the exhaust stream. This fluid reacts with the NOx emissions in the presence of a catalytic converter to convert harmful NOx into harmless nitrogen gas N_2 and H_2O (Figure S.3).

FIGURE S.3 Schematic representation of primary after-treatment system [2].

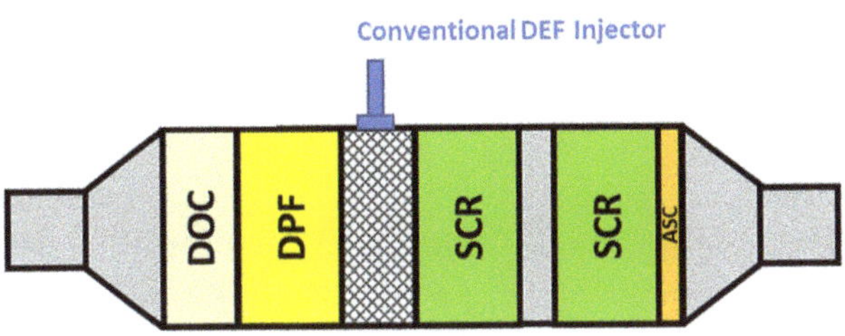

An SCR system injects a urea-based solution (DEF) into the exhaust stream upstream of a catalytic converter. The high-temperature exhaust gases cause DEF to decompose into ammonia (NH_3) and CO_2. NH_3 from the DEF reacts with NOx in the presence of a catalytic converter, breaking them down into N_2 and H_2O. SCR technology significantly reduces NOx emissions, which are harmful pollutants contributing to air pollution and smog formation. The catalytic converter within the SCR system contains catalysts that facilitate the chemical reactions between DEF, NH_3, and NOx. Sensors monitor exhaust gases and DEF injection rates to ensure optimal performance and emission reduction.

Secondary Air

Secondary air brake pressure refers to the air pressure level within the secondary air brake system of a commercial vehicle equipped with an air brake system. Air brakes are commonly used in large trucks, buses, and other heavy-duty vehicles to provide effective braking performance.

Commercial vehicles with air brake systems often have dual or split braking systems: the primary (or service) and the secondary (or emergency) systems. The secondary air brake system is designed as a backup or emergency system to ensure braking capability even if the primary system fails or loses pressure. In an air brake system, braking is achieved by applying air pressure to brake chambers, which actuate the brakes on the vehicle's wheels.

The secondary air brake pressure is the amount stored in the secondary system's air reservoirs. This pressure is used to apply emergency braking in situations where the primary system may be compromised. When the secondary air brake system is activated, the stored air pressure is utilized to apply braking force to the vehicle's wheels, allowing the driver to bring the vehicle to a stop in emergencies.

For more information, see **FMVSS 121** [3].

Section Control

GPS-based technology automatically controls the individual sections of planting equipment to prevent double planting or overlapping. Section control is a precision farming technique and technology used in agriculture to control the respective sections of planting, seeding, spraying, or fertilizing equipment to avoid overlapping or double application. It involves the automatic or manual adjustment of machinery sections to ensure that each part of a field is treated only once, reducing waste of resources, such as seeds, chemicals, and fuel, while optimizing crop production and minimizing environmental impact. Section control is often implemented using GPS technology and specialized software to manage farming equipment in the field accurately.

Seed Drill

A seed drill is an agricultural implement for planting seeds in rows at a consistent depth and spacing in a prepared seedbed or field. It is not a commercial vehicle but rather a piece of machinery that can be attached to a tractor or pulled by draft animals. The seed drill consists of a hopper for holding seeds, a mechanism for regularly metering and delivering seeds to the ground, and devices to cover the seeds with soil.

The primary purpose of a seed drill is to improve the efficiency and accuracy of seed placement in the soil, which results in better seed germination rates and crop uniformity. It is critical in modern farming practices, enabling farmers to sow seeds with precision, reduce seed wastage, and maximize crop yields.

Self-Propelled Combine Harvester

A self-propelled combine harvester is a specialized agricultural commercial vehicle designed for the mechanized harvesting of crops, particularly grains such as wheat, barley, corn, and rice. Unlike traditional combine harvesters that may require external power sources, a self-propelled combine harvester integrates its propulsion system, allowing it to move independently through fields during harvesting operations.

A self-propelled combine harvester features an onboard propulsion system, typically powered by an ICE and batteries [4], allowing it to move autonomously across fields during harvesting. Equipped with a cutting mechanism, threshing unit, and cleaning system, the combine harvester efficiently harvests, separates and processes crops in a single pass. The harvested crop is then collected and conveyed into a storage bin. The harvested grain is stored in an onboard grain tank or hopper, minimizing the need for frequent unloading and maximizing harvesting efficiency.

For more information, see **J1614** [5].

Self-Propelled Sprayer

A self-propelled sprayer is a specialized commercial vehicle that applies agricultural chemicals, including pesticides, herbicides, fertilizers, and other crop protection or nutrient solutions. Unlike traditional towed or tractor-mounted sprayers, a self-propelled sprayer integrates its propulsion system [4], allowing it to move independently through fields during spraying operations.

For more information, see **J1614** [5].

Semi-Trailer

A semi-trailer refers to the rear portion of a tractor–trailer combination used in heavy-duty transportation. A semi-trailer is a specialized cargo container towed by a tractor (cab or truck) and designed to carry various goods. It is a critical component of the tractor–trailer configuration, allowing for efficient and versatile cargo transportation.

The semi-trailer is a large container-like structure designed to carry goods of different sizes, shapes, and types. A semi-trailer is connected to the tractor's fifth-wheel coupling, allowing it to pivot and turn independently while being towed. Semi-trailers have wheels and axles to support the cargo's weight.

Semi-trailers come in various designs to accommodate different types of cargo, including dry goods, refrigerated items, liquids, and bulk materials. Semi-trailers can vary in length, width, and height to adhere to legal weight and size limits. Semi-trailers are equipped with landing gear, including adjustable legs that offer support when the trailer is not attached to the tractor.

Semi-Truck

A semi-truck, a tractor-trailer or 18-wheeler, is a heavy-duty commercial vehicle designed to transport large quantities of goods over long distances. A semi-truck consists of two main components: the tractor (the cab or truck) and the trailer. The tractor provides the power to propel the vehicle and is connected to the cargo trailer.

The front portion of a semi-truck is the tractor, which houses the driver's compartment, engine, transmission, and various control systems. The rear portion of a semi-truck is the trailer, a large, specialized container designed to carry various types of cargo. The tractor and the trailer are connected through a fifth-wheel coupling, allowing them to pivot and turn independently.

Many semi-trucks have tandem axles (two sets of axles) on the tractor and trailer, which help distribute the weight and improve stability. Semi-trucks are typically powered by powerful diesel engines that provide the necessary torque for heavy loads and long distances. Aerodynamic design elements are often incorporated to reduce air resistance and improve fuel efficiency.

Service Truck

A service truck is a specialized vehicle with tools, equipment, and supplies to provide on-site maintenance, repairs, and services to other vehicles, machinery, or facilities. Service trucks are commonly used in construction, utilities, towing, and emergency response industries, enabling technicians and service personnel to address issues quickly and efficiently.

A service truck is typically equipped with a utility bed or compartmentalized body that contains storage space for tools, equipment, and supplies. The utility bed features compartments, drawers, racks, and shelving to organize and store various tools and equipment. Service trucks may carry wrenches, sockets, power tools, welding equipment, hydraulic jacks, air compressors, and more, specific to their services. Some service trucks are equipped with generators or power inverters to provide electrical power for tools and equipment.

Depending on the nature of the services offered, a service truck might have a built-in crane or lift mechanism to handle heavy loads or machinery. Service trucks can deliver a range of capabilities, from providing roadside assistance and repairs to conducting maintenance and equipment installations. Service trucks can be customized to suit the needs of different industries, such as plumbing, electrical, construction, or emergency response.

Shuttle Bus

Shuttle buses refer to compact to mid-sized commercial vehicles designed for short-distance transportation of passengers within defined routes or areas. Shuttle buses are used to provide convenient and efficient transportation services.

Shuttle buses come in various sizes with different seating capacities to accommodate multiple passengers, from smaller groups to larger crowds. Shuttle buses typically have a compact or mid-sized design, making them suitable for navigating through urban streets, campuses, airports, and other areas with tight spaces. Many shuttle buses have numerous entry and exit doors to facilitate quick boarding and alighting. These buses are designed for short-distance travel within specific routes, areas, or facilities. Some shuttle buses are created with features such as low floors, ramps, and lifts to ensure accessibility for passengers with disabilities.

Skid Steer Loader

A skid steer loader is a compact and versatile commercial vehicle characterized by its small size, rigid frame, and distinctive lifting arms. This loader type is designed for various material handling and construction tasks. It is known for its ability to pivot or "skid" by independently controlling the speed of its wheels or tracks on either side. Skid steer loaders are used in construction, landscaping, agriculture, and other industries for digging, lifting, grading, and transporting materials.

Sleeper

A sleeper refers to the integrated sleeping compartment found in long-haul trucks and some other types of commercial vehicles. The sleeper provides resting and sleeping accommodations for drivers during extended trips, allowing them to comply with HOS regulations and rest comfortably between driving shifts.

A sleeper typically features a bed, mattress, and bedding to provide a comfortable sleeping space for the driver. Sleeper compartments often include storage cabinets, shelves, and compartments for personal belongings, clothing, and essentials. Some sleepers have basic amenities, such as a small desk, power outlets, lighting, and ventilation. The sleeper compartment offers a private space for drivers to rest and relax without leaving the vehicle.

Long-haul truck drivers are subject to HOS regulations dictating maximum driving and duty periods. Sleepers allow drivers to rest during mandatory off-duty periods. Sleepers improve driver comfort during extended trips, reducing the need for frequent overnight hotel stays.

The most common type of sleeper features a single or double bed arrangement above the cab area. Mid-roof sleepers have a lower profile and offer more storage space but may have limited headroom. High-roof sleepers provide more vertical space, allowing drivers to stand upright inside. They deliver enhanced comfort for longer trips. Some commercial vehicles can be equipped with custom sleepers, additional amenities, and comfort features.

Slip

A slip refers to the loss of traction between a vehicle's tires and the road surface, reducing control and the potential for skidding or sliding. Slipping occurs when the friction between the tires and the road is insufficient to maintain proper traction, often due to adverse weather conditions, road surface conditions, or excessive speed [6].

Rain, snow, ice, or slush on the road surface can reduce tire traction and lead to slipping. Spilled oil, grease, or other slippery substances on the road can create hazardous conditions and cause vehicles to slip. Driving at speeds too high for the road conditions can cause tires to lose traction and slip, especially during turns or sudden maneuvers. Worn-out or improperly inflated tires may have reduced grip on the road, increasing the likelihood of slipping.

Abrupt braking or rapid acceleration can overwhelm tire grip and lead to slipping. Vehicles without proper traction aids, such as snow tires or tire chains in snowy conditions, are more prone to slipping.

Slipping reduces a driver's ability to control the vehicle's direction and speed, increasing the risk of accidents. Slipping can lead to skids or slides, where the vehicle's tires lose lateral or longitudinal traction and move in an unintended direction. Uncontrolled slipping can result in collisions with other vehicles, obstacles, or barriers. Slipping can lead to loss of vehicle stability, making it challenging to recover control.

Snowplow

A snowplow is a specialized attachment or implement designed to clear snow and ice from roads, streets, parking lots, and other surfaces. Snowplows are commonly mounted on the front of trucks, tractors, or other vehicles, and they are used to push, lift, or throw snow out of the way to maintain safe and passable road conditions during winter weather.

A snowplow is equipped with a large, flat, and often adjustable blade that comes into contact with the snow-covered surface. The primary function of a snowplow is to remove accumulated snow and ice from roadways and surfaces, allowing vehicles and pedestrians to navigate safely. Many snowplow blades can be angled to the left or right, permitting the operator to push snow to the side as they move forward. Snowplows have a lifting mechanism that raises or lowers the blade, enabling operators to adjust the blade's height and engage or disengage it. Some snowplows have additional features, such as salt or sand spreaders, to enhance traction on icy surfaces.

There are a few types of snowplows. The straight-blade plow, the most common snowplow, features an adjustable straight blade that can be angled to the side. The V-plows have a V-shaped blade that can be opened or closed to break through heavy snow effectively and direct it to the sides. Wing plows have multiple sections that can be adjusted to create a wider plowing path. A scoop plow has a curved design that helps scoop up and throw snow to the side. Pusher plows are typically used for larger areas and push snow to the sides without lifting it.

Sprayer

Sprayers refer to specialized vehicles or equipment that apply liquids, such as pesticides, herbicides, fertilizers, or water-based solutions, onto crops, plants, or surfaces. These vehicles play a crucial role in agricultural, horticultural, and industrial applications for the controlled and efficient distribution of liquids.

Sprayers are equipped with tanks that hold the liquid solution to be sprayed. Tank sizes can vary depending on the vehicle's intended use and capacity requirements. Sprayers use various methods to distribute the liquid, such as nozzles, booms, and wands, allowing for precision in application. Sprayers have pump systems that pressurize the fluid and facilitate its even distribution through spraying.

Many modern sprayers have advanced control systems that allow operators to adjust spray pressure, coverage, and other parameters. Agricultural sprayers often have booms, extending arms equipped with multiple nozzles to cover a broader area in a single pass.

Spring Brake Chamber (Air Brakes)

The spring brake chamber is a combination service and spring brake chamber assembly for truck, tractor, trailer, and bus applications. The design basically provides two individual chambers in one housing with their own diaphragms, springs, and separate air supplies. Spring brake chambers provide a safe parking brake arrangement when system air pressure is exhausted from the spring brake chamber. A large coil spring then forces the pushrod forward to apply the brakes mechanically. For the combination service and spring parking brake systems to operate independently of each other, both service and parking brake air lines must be used. The service line transfers air from the brake pedal only to the service chamber portion of the assembly. The parking airline is connected to the dash-mounted spring brake control valve button and can direct air only into the spring brake chamber portion of the assembly. Spring brakes are either sealed or clamp-type designs. Those that are attached to the rear of the service brake chamber are referred to as piggyback spring brakes.

For more information on *Braking System Definitions—Truck and Bus*, see **J2627_200908** [7].

Spring Brake Control Valve (Inversion Valve Air Brakes)

The spring brake control valve supplies air to the hold-off chambers to release the spring brakes and enables the vehicle to maneuver. During normal operation, it limits the hold-off pressure to the spring brake chambers to worth around 90 psi. This action speeds application times in an emergency and permits a uniform hold-off pressure (system pressure fluctuates between cut-in/cut-out pressure values). Some systems use an inversion valve. In the event of a primary air system failure, the inversion valve allows for a modulated application of the emergency air brake chambers by relieving air from the hold-off chamber in direct proportion to the applied pressure at the treadle valve. Spring brake control valves are often incorporated into multifunction valves that contain parking, service, and inversion functions. The trailer air system has its own spring brake control valve, which is intended to:

- Automatically apply trailer spring brakes in the event of a breakaway or trailer itinerary failure.
- Protect trailer reservoir(s) gas pressure in the event of a breakaway or trailer supply route failure.
- Prevent an automatic trailer spring brake application with atmospheric pressure loss within the trailer reservoir(s).
- Offer an optional anti-compounding feature and stop overriding the service brake signal while the trailer supply route is at gas pressure (no pressurized air in line).
- Prevent charging of the trailer system should a failure occur within the emergency brake circuit.

For more information on *Air Brake Valve – Performance Requirements*, see **J1410_202012** [8].

Springs (Automotive Suspension)

Springs are located between the frame and the axle assemblies and are shaped to suit specific applications. Leaf springs are normally semi-elliptical. They absorb the applied force (pressure of the load) by flattening out under the load. Springs are used on the underside of a car or truck to help the vehicle carry larger loads. Coil springs are formed in an exceedingly spiral from one steel rod. They absorb the force of impact by twisting and compressing. They are used on smaller vehicles to smooth the ride and improve handling. Torsion bars are held rigidly at one end and denote their center because the control arm is deflected. Control arms, also called A-arms or wishbones, attach to the chassis with rubber bushings and permit the movement of the tire and wheel assembly. Torsion bars all return to their original shape when the deflecting force (a force that moves an object in a very different direction or into a special shape) is removed. They are typically employed within the front of most pickup trucks because they handle better than leaf springs.

For information on helical compression and extension spring terminology, see **J1121_201608** [9].

Spur Gear

Gears with straight teeth are mounted on parallel shafts. You will generally only find spur gears on heavy transmissions used in medium- and heavy-duty trucks and construction equipment. Each time a gear tooth engages a tooth on the other gear, the teeth collide, and this makes noise. It also increases stress on the gear teeth. To reduce the noise and stress in the gears, most automotive gears are helical.

Square Baler

A square baler is a specialized agricultural implement towed or mounted on a commercial tractor, forming an integral part of the mechanized harvesting process. Its primary function is compressing and bundling hay, straw, or other crop residues into compact and stackable rectangular bales. The square baler is instrumental in creating a uniform, easy-to-handle bales for storage, transportation, and subsequent agricultural applications.

Square balers utilize a mechanical system to pick up, compress, and bind cut crop material into rectangular bales. This process involves a series of moving components, including a pickup mechanism, feeder, plunger, and knotter or twine system. Unlike round balers, which produce cylindrical bales, square balers create rectangular bales with flat sides and edges. The shape facilitates efficient stacking and handling. Equipped with a pickup system, the square baler efficiently gathers cut crop material from the field, ensuring a continuous and uniform flow into the baling chamber. The compression chamber is a crucial component where the gathered crop material is compacted into a dense and rectangular form. The plunger, powered by the tractor's PTO, plays a crucial role in this compression process. A knotter or twine system secures the compressed material, forming a tightly bound rectangular bale. This ensures the integrity and stability of the bale during handling, storage, and transportation. Many square balers offer adjustable settings to control the dimensions and density of the produced bales. This flexibility allows farmers to tailor the bale size based on specific storage and handling requirements.

Square balers are typically connected to the rear of a commercial tractor, relying on the tractor's power and hydraulics for operation. The tractor's PTO provides the power for the baler's mechanical components.

Star Wheel

A toothed wheel adjusting screw is used to adjust drum brakes or a truck clutch.

Starter Motor

A starter motor is an essential automotive component responsible for initiating the ICE's operation. The starter motor uses the engine's flywheel or flexplate to crank the engine and enable the combustion process, allowing the engine to start running.

Steer-by-Wire System

A steer-by-wire system is an advanced technology that replaces traditional mechanical steering components with electronic and electromechanical components, such as the steering column, steering shaft, and hydraulic power assist. This system allows for the control of the vehicle's steering purely through electronic signals and sensors, enhancing vehicle maneuverability, driver assistance, and integration with other vehicle systems.

Steering System

A steering system refers to the combination of components and mechanisms that allow the driver to control the direction of the vehicle's movement. The steering system enables the driver to maneuver the vehicle, change its course, and navigate safely on roads and highways.

The steering wheel is the primary control interface that the driver uses to turn the vehicle [10]. It is connected to the steering column, which transfers the driver's input to the rest of the steering system. The steering gear mechanism converts the rotational motion of the steering wheel into the lateral movement needed to turn the vehicle's wheels. The steering linkage consists of various rods, arms, and joints that transmit the steering motion from the steering gear to the wheels.

The pitman's arm is part of the steering linkage that connects the steering gear to the wheels. It translates the rotational motion of the steering gear into lateral movement. Tie rods are components of the steering linkage that connect the pitman arm or center link to the steering knuckles of the front wheels. Steering knuckles are the parts to which the front wheels are attached. They pivot to allow the wheels to turn left or right. Many commercial vehicles are equipped with power steering systems that assist the driver in turning the wheels by using hydraulic or electric power to reduce the effort required.

Commonly used in smaller vehicles, this system utilizes a rack-and-pinion mechanism to translate steering wheel rotation into lateral movement. The recirculating ball steering is found in larger vehicles; a system uses a recirculating ball mechanism to transfer the steering motion.

The integral power system integrates the power steering assist mechanism with the steering gear, enhancing the ease of steering. Electric power steering systems use an electric motor to provide steering assistance, offering efficiency and adaptability. Hydraulic power steering systems use hydraulic pressure to assist steering, enhancing driver comfort.

Step Deck Trailer

A step deck trailer, also known as a drop deck trailer or lowboy trailer, refers to a flatbed trailer with two distinct deck levels: an upper deck and a lower deck. The lower deck is positioned closer to the ground than the upper deck, allowing for taller cargo to be transported while complying with height restrictions.

A step deck trailer is characterized by its dual-deck design, with a higher deck at the front (near the truck) and a lower deck at the rear. This design accommodates taller cargo. The lower deck provides clearance for taller loads that may exceed the height limitations of traditional flatbed trailers, making it suitable for transporting equipment, machinery, and other oversized items. Step deck trailers can haul a wide range of cargo, including machinery, construction equipment, vehicles, lumber, and other items exceeding the height of standard flatbed trailers. The lower deck's positioning contributes to a lower load height compared to traditional flatbed trailers, which can simplify loading and unloading operations.

Some step deck trailers are equipped with ramps or a beavertail at the rear to facilitate the loading and unloading of heavy equipment and vehicles. Step deck trailers are designed to distribute the load weight more evenly across their axles and provide stability during transport. Step deck trailers help comply with height restrictions imposed by transportation regulations, allowing the transportation of taller cargo within legal limits.

Step Van

A step van is a type of delivery or utility vehicle characterized by its boxy and compact design. Step vans are often used for delivering packages, goods, and services, especially in urban and suburban settings. They are designed for efficient loading and unloading, providing easy access to the cargo area from the driver's compartment.

Step vans typically have a box-like body design, offering a spacious and enclosed cargo area for transporting goods. The low floor height of a step van allows for easy entry and exit for the driver and delivery personnel without the need for stairs or ladders. Step vans usually have a large sliding door on the side, enabling quick and convenient access to the cargo compartment. The rear of the step van is equipped with double doors that swing open or slide apart, providing a wide opening for loading and unloading. Step vans are designed to navigate urban environments efficiently, making them suitable for delivery routes in crowded areas.

Step vans can be customized to suit specific business needs, with shelving, racks, or compartments added to the cargo area for organized storage. Beyond package delivery, step vans are used for various purposes, including mobile workshops, food trucks, mobile retail, and service vehicles.

Sugarcane Harvester

A sugarcane harvester is a specialized commercial vehicle designed for the efficient mechanized harvesting of sugarcane crops in agricultural settings (Figure S.4). This heavy-duty machine is equipped with cutting and handling mechanisms to streamline the process of harvesting, transporting, and loading sugarcane for further processing in the sugar industry.

FIGURE S.4 Sugarcane harvester in the field.

Sugarcane harvesters are equipped with cutting mechanisms, such as sharp blades or discs, designed to sever the sugarcane stalks at the base efficiently during the harvesting process. The harvester incorporates a gathering system that collects the cut sugarcane stalks and conveys them through the machine for further processing. This system ensures a continuous and uniform flow of sugarcane. As the sugarcane moves through the harvester, cleaning mechanisms remove excess leaves and debris, leaving only the clean sugarcane stalks. Detaching devices may also be employed to separate the sugarcane from the soil. Sugarcane harvesters often feature chopping mechanisms that cut the harvested sugarcane into manageable lengths. The chopped sugarcane is then loaded onto a transport conveyor for efficient transportation to collection points or processing facilities. Many sugarcane harvesters are equipped with adjustable cutting height settings, allowing farmers to optimize the harvesting process based on sugarcane field conditions and crop maturity.

Sugarcane harvesters are designed with an operator cabin that provides a comfortable and ergonomic workspace. The cabin typically includes controls for monitoring and adjusting various harvesting parameters. Sugarcane harvesters are often mounted on tracks or wheels, providing the mobility necessary to navigate through sugarcane fields. This ensures accessibility and efficiency in harvesting operations. Some sugarcane harvesters are designed to integrate with transport vehicles, facilitating the direct loading of harvested sugarcane onto trailers for transportation to processing facilities.

Suspension

Suspension refers to the system of components that connect a vehicle's chassis or body to its wheels, allowing the wheels to move independently of its frame. The primary purpose of a suspension system is to provide a comfortable ride for occupants, maintain proper wheel contact with the road, and absorb shocks and vibrations caused by uneven road surfaces.

The suspension system plays a critical role in providing a smooth and comfortable ride for passengers and cargo by minimizing the effects of road irregularities and vibrations. A well-designed suspension system enhances a vehicle's handling and stability by maintaining consistent tire contact with the road, improving steering response, and reducing body roll during cornering. The suspension system helps ensure the vehicle's tires maintain adequate traction, particularly during acceleration, braking, and turns. The suspension absorbs shocks and impacts from road imperfections, such as potholes, speed bumps, and rough surfaces, preventing these forces from directly affecting the vehicle's occupants and cargo.

There are several types of suspension systems. Each wheel moves independently in an independent suspension system, providing better comfort and handling. In solid axle suspension, both wheels on an axle move together, making it suitable for heavy-duty applications. Air suspensions use air springs for adjustable ride height and comfort, which is common in buses and trucks. Leaf spring suspension uses stacked metal leaves for support, commonly found in trucks and vans.

For more information, see **J1574** [11].

REFERENCES

1. Society of Automotive Engineers, "J284 Safety Alert Symbol for Agricultural, Construction, and Industrial Equipment," SAE Publishing, Warrendale, 2021.
2. Meruva, P., Matheaus, A., Zavala, B., Sharp, C.A. et al., "Application of Model-Based Controller on a Heavy-Duty Dual Selective Catalytic Reduction Aftertreatment," *SAE Int. J. Engines* 16, no. 5 (2023): 681-691, doi:https://doi.org/10.4271/03-16-05-0040.
3. US Department of Transportation, "Federal Motor Vehicle Safety Standard 121 Warning Signals, Air Pressure and Vacuum Gauges," National Highway Traffic Safety Administration, Washington DC, 2005.
4. Society of Automotive Engineers, "J930 Storage Batteries for Off-Road Self-Propelled Work Machines," SAE Publishing, Warrendale, 2022.
5. Society of Automotive Engineers, "J1614 Wiring Distribution Systems for Off-Road, Self-Propelled Work Machines," SAE Publishing, Warrendale, 2018.
6. Society of Automotive Engineers, "J2115 Air Brake Performance and Wear Test Code Commercial Vehicle Inertia Dynamometer," SAE Publishing, Warrendale, 2020.
7. Society of Automotive Engineers, "Standard J2627_200908: Braking System," SAE Publishing, Warrendale, 2009.
8. Society of Automotive Engineers, "Standard J1410_202012: Air Brake Valve -," SAE Publishing, Warrendale, 2020.
9. Society of Automotive Engineers, "Standard J1121_201608: Helical," SAE Publishing, Warrendale, 2016.
10. Society of Automotive Engineers, "J2425 Steering Control Systems - Heavy Trucks," SAE Publishing, Warrendale, 2021.
11. Society of Automotive Engineers, "J1574 Measurement of Vehicle and Suspension Parameters for Directional Control Studies," SAE Publishing, Warrendale, 2018.

T

Tag Axle

A tag axle is an additional axle installed on a vehicle behind the drive axle to provide extra support and distribute weight, mainly in heavy-duty trucks, buses, and other large vehicles. Unlike the primary drive or steer axles, a tag axle does not power the wheels or participate in steering the vehicle. Instead, its primary purpose is to increase load-bearing capacity and enhance stability, especially when carrying heavy loads.

A tag axle helps distribute the weight of the vehicle and its cargo over a larger area, reducing the load on individual axles and tires. This is especially important when carrying heavy or bulky loads. Some tag axles can be lifted off the ground when unnecessary, reducing tire wear, fuel consumption, and maneuvering restrictions when the extra axle is unnecessary.

The presence of a tag axle helps improve stability during cornering and reduces the risk of weight shifts that can lead to rollovers. In many jurisdictions, commercial vehicles have weight limits per axle. A tag axle helps ensure compliance with weight regulations by distributing the load more evenly.

Tandem Drive Axle

A heavy-truck rear final drive assembly that combines two single-axle assemblies with a power divider, connected via a short driveshaft.

Tank Trailer

A tank trailer is a specialized type designed to transport large quantities of liquids, gases, or bulk materials. Like a tank truck, a tank trailer has a cylindrical tank or container securely mounted on the trailer chassis. Tank trailers are commonly used in various industries to haul multiple substances, including petroleum, chemicals, food, and hazardous materials.

Tank trailers are built with cylindrical tanks of materials suitable for transporting cargo. These tanks can be made from aluminum, stainless steel, carbon steel, or other appropriate materials. Tank trailers come in various sizes and capacities, with tanks capable of holding varying volumes of liquids or bulk materials, often measured in gallons or cubic feet. Depending on the type of cargo, tank trailers can have single-compartment designs for carrying a single substance or multi-compartment designs for transporting different substances simultaneously.

Tank trailers for transporting hazardous materials adhere to stringent safety regulations and standards to prevent leaks, spills, or contamination. Tank trailers are equipped with loading and unloading systems, including pumps, hoses, and fittings that connect to loading or unloading stations. Tank trailers are designed to safely transport a diverse range of cargo, from liquids like gasoline, chemicals, and food products to bulk materials like cement or sand.

Tank Trucks (Fuel, Water, and Chemicals)

A tank truck is a specialized motorized vehicle designed to transport liquids, gases, or bulk materials in large tanks mounted on the chassis. These trucks are commonly used to haul a wide range of substances, including liquids such as water, chemicals, fuels, and gases, in bulk quantities from one location to another.

Tank trucks have tanks of varying shapes and sizes, depending on the type of cargo they carry. Tanks can be cylindrical, oval, or customized to fit specific requirements. Tanks are typically constructed from materials suitable for the cargo being transported, such as aluminum, steel, or specialized materials for hazardous substances. Some tank trucks have multiple compartments within the tank to haul different types of liquids or materials simultaneously. Tank trucks are equipped with pumping systems to load and unload cargo efficiently. Tank trucks include safety features like baffles within the tank to prevent liquid sloshing, vapor recovery systems, and spill containment measures.

Tank trucks are used across various industries for transporting different types of substances. For example, bulk liquids: tank trucks transport water, chemicals, milk, edible oils, and beverages. Tank trucks deliver gasoline, diesel, and aviation fuel to gas stations, airports, and industrial facilities. These trucks transport chemicals used in manufacturing, agriculture, and other industries. Tank trucks carry gases, such as propane, oxygen, nitrogen, and liquefied petroleum gas (LPG), in pressurized or cryogenic tanks.

Tare Weight

Tare weight refers to the weight of a vehicle or container when it is empty or unloaded. It represents the weight of the vehicle's structure, chassis, body, and any other components, excluding any cargo, passengers, fuel, or other items that might be carried.

Tare weight is an essential measurement for various reasons. Tare weight is used as a baseline for calculating the maximum payload capacity of a vehicle. Subtracting the tare weight from the GVWR gives the maximum weight of cargo and passengers that the vehicle can safely carry. Knowing the tare weight helps ensure the load is appropriately distributed within the vehicle to maintain balance and stability. Tare weight is crucial for complying with weight regulations imposed by transportation authorities. Exceeding weight limits can lead to fines and safety concerns. A lighter tare weight can contribute to better fuel efficiency, as the vehicle requires less energy to move its weight. Tare weight calculates the weight distribution on each axle, helping prevent overloading and maintain safe handling. Tare weight is a factor in determining appropriate tire pressure and suspension settings, which can affect vehicle performance and longevity.

Tare weight can be determined through various methods. Vehicle manufacturers provide tare weight information for their models as part of the specifications. Tare weight can be measured using weighing scales designed to handle the entire vehicle's weight. If the GVW is known (weight when loaded) and the payload is empty, tare weight can be calculated by subtracting the payload weight from the GVW.

Taxi

A taxi is a licensed passenger vehicle that offers transportation services to the public for a fare. Taxis are commonly used for point-to-point travel within urban areas, providing convenient and on-demand transportation for individuals or small groups.

Licensed taxis are operated by certified drivers and companies that comply with local regulations and transportation authorities. Metered taxis typically use a metered fare system that calculates the cost of the ride based on factors such as distance traveled and time spent. Dedicated signage taxis often have a distinctive design and signage that makes them easily recognizable, helping passengers identify them as available for hire.

Telematics

See **Telematics System** and **GPS**.

Telematics System and GPS (Global Positioning System)

Telematics refers to integrating telecommunications and information technology in vehicles to enable the collection, transmission, and analysis of data related to vehicle operations and driver behavior. Telematics systems utilize various technologies to gather and communicate data, offering insights into vehicle performance, location, safety, and more.

GPS is a satellite-based navigation and positioning system. GPS utilizes a network of satellites to accurately determine the geographic location, speed, and altitude of a vehicle or device equipped with a GPS receiver.

GPS provides real-time and accurate information about the vehicle's location, helping with route optimization, navigation, and asset tracking. GPS offers turn-by-turn directions and route guidance, improving driver efficiency and reducing the likelihood of getting lost. Telematics systems, often integrated with GPS, enable fleet managers to monitor multiple vehicles' location, status, and performance within a fleet. Telematics systems can provide alerts for unsafe driving behaviors, such as speeding or sudden braking, improving driver safety.

Telematics systems and GPS offer insights into driver behavior, fuel consumption, and vehicle utilization, allowing for the optimization of routes and operations. Telematics systems help track valuable assets and cargo, enhancing security and theft prevention. Telematics systems can transmit data related to vehicle diagnostics, allowing for predictive maintenance and preparation, and minimizing downtime.

A telematics system and GPS typically include several components. The GPS receiver receives signals from satellites to determine the vehicle's location. The telematics unit processes GPS data, sensor data, and other vehicle information and transmits it using wireless communication. Telematics systems use cellular networks, satellite communication, or other wireless technologies to send data to a central monitoring station. The central monitoring station receives, processes, and displays data from multiple vehicles, allowing fleet managers to monitor and manage the fleet. Fleet managers and operators access data and insights through software applications or Web interfaces. Telematics systems may incorporate sensors to monitor vehicle parameters like speed, fuel consumption, and engine diagnostics.

Telemetry System

Telemetry systems are technologies and devices that remotely collect and transmit data from vehicles to a central monitoring station or control center. These systems enable real-time monitoring and communication of vehicle parameters and performance metrics.

Telemetry systems gather data from various sensors and components within a commercial vehicle, including engine parameters, vehicle speed, location, fuel consumption, tire pressure, temperature, and more. Telemetry systems provide real-time visibility into the vehicle's status, allowing fleet managers or operators to monitor vehicle performance and conditions remotely. Telemetry systems can transmit DTCs and alerts related to engine malfunctions, maintenance requirements, and potential issues. Telemetry systems offer insights into driver behavior, vehicle utilization, and fuel consumption, enabling optimization of routes, scheduling, and operations for improved efficiency.

Telemetry systems can provide alerts for unsafe driving behavior, such as excessive speeding, harsh braking, or sudden lane changes, contributing to improved driver safety. Telemetry systems enable fleet managers to track multiple vehicles' location, status, and performance within a fleet. Telemetry systems allow original equipment manufacturers (OEMs) to alter the software components of the vehicle without bringing the vehicle into a servicing center for product updates. Some advanced telemetry systems enable remote control of vehicle functions, such as engine shutdown, ignition control, or unlocking doors.

A telemetry system typically consists of several components. Sensors collect data from various vehicle components, including the engine, transmission, brakes, and tires. The telematics unit processes and aggregates sensor data and transmits it to the central monitoring station using wireless communication technologies (e.g., cellular networks and satellite communication). The central monitoring station (back office) receives and processes data from multiple vehicles, allowing operators to monitor and manage the entire fleet (**Figure T.1**). Telemetry systems utilize cellular networks, satellite communication, or other wireless technologies to transmit data. Fleet managers and operators access data and insights through software applications or Web interfaces.

FIGURE T.1 The telematics system from vehicle, back office (cloud) and remote (mobile) location [1].

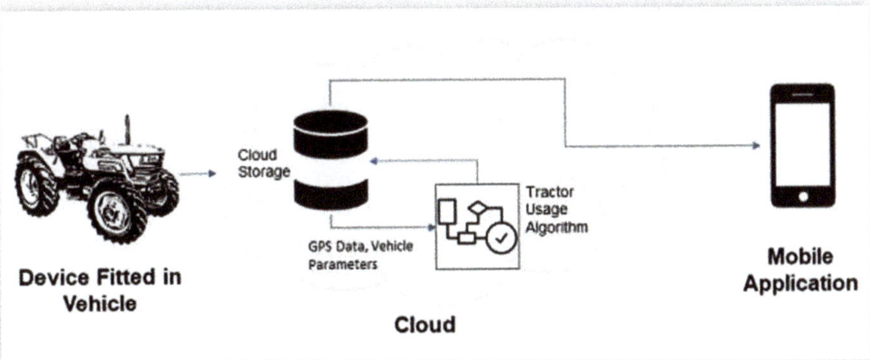

Tire

Tires refer to the rubber components mounted on the wheels of a vehicle that provide contact with the road surface. Tires play a critical role in supporting the vehicle's weight, transmitting traction and braking forces, and contributing to overall vehicle performance and safety.

Tires bear the vehicle's and its cargo's weight, distributing the load evenly across the road surface. Tires provide grip and traction between the vehicle and the road, enabling acceleration, deceleration, and cornering. Tires play a vital role in braking by converting kinetic energy into heat through friction between the tire and the road surface. Tires act as a cushion, absorbing shocks and vibrations from the road, which improves ride comfort for occupants and helps protect the vehicle's components.

Tires contribute to steering responsiveness and stability, allowing the driver to control the direction of the vehicle. Properly inflated and well-maintained tires contribute to better fuel efficiency by reducing rolling resistance.

Tire Pressure Control System

A tire pressure control system refers to an advanced electronic system designed to actively and automatically adjust the air pressure in a vehicle's tires based on various factors such as load, road conditions, and driving speed. The primary purpose of a tire pressure control system is to maintain optimal tire pressure for improved safety, performance, and efficiency [2].

A tire pressure control system continuously monitors the air pressure in each tire and adjusts it as needed to maintain the recommended pressure levels. Some systems can detect changes in the vehicle's load and adjust tire pressure accordingly to ensure optimal contact with the road surface. Certain tire pressure control systems can adapt tire pressure based on road conditions, such as soft off-road terrain or highway driving.

Maintaining the correct tire pressure contributes to better fuel efficiency. A tire pressure control system helps achieve consistent fuel economy by ensuring optimal tire inflation. Proper tire pressure is crucial for maintaining traction and stability, especially in varying road and weather conditions. Consistently maintaining optimal tire pressure helps extend tire life by promoting even wear. A tire pressure control system eliminates the need for manual pressure checks and adjustments, enhancing driver convenience.

A tire pressure control system typically consists of a few components. Pressure sensors are installed inside each tire to continuously measure air pressure and transmit data to the control unit. The control unit processes sensor data, assesses the vehicle's conditions, and determines the appropriate tire pressure adjustments. Some tire pressure control systems have an air compressor that can inflate or deflate tires as needed. The system's valves regulate airflow into and out of the tires, allowing pressure adjustments. Many systems feature a display on the dashboard that shows the real-time tire pressure of each tire.

TMC (Technology Maintenance Council)

The TMC is an industry organization and council that operates within the ATA. The TMC focuses on advancing technology and maintenance practices in the commercial vehicle and transportation industry. It serves as a platform for professionals, including fleet maintenance personnel, engineers, technicians, and industry suppliers, to collaborate, share knowledge, and address challenges related to vehicle maintenance, technology integration, and equipment reliability.

The TMC provides a collaborative forum for professionals within the commercial vehicle industry to share insights, expertise, and best practices related to technology, maintenance, and equipment reliability. The TMC organizes educational programs, seminars, and workshops to enhance the skills and knowledge of industry professionals in areas such as vehicle maintenance, diagnostics, and the integration of advanced technologies. The TMC establishes task forces and committees focusing on specific areas of interest, such as fuel efficiency, vehicle safety, telematics, and emerging technologies. These groups work collaboratively to develop guidelines and recommendations for industry practices. TMC often features a technology showcase where companies can exhibit the latest advancements in commercial vehicle technology, maintenance tools, and related products.

TMS (Transportation Management System)

TMS is a software solution that assists in planning, executing, and optimizing transportation operations. It covers route planning, load optimization, carrier selection, and order management.

Tone Ring

A tone ring refers to a component used in automotive and commercial vehicle systems, particularly in ABS and traction control systems. The tone ring is a toothed or notched ring made of metal or other magnetic material that is attached to a rotating component, such as a wheel hub or axle. The primary function of the tone ring is to provide a signal to sensors, allowing the vehicle's electronic systems to monitor the rotational speed of the component.

Tone rings are commonly used in ABS to monitor the rotational speed of each wheel. By measuring the speed of individual wheels, the ABS can detect and prevent wheel lockup during braking, enhancing stability and control. Tone rings also play a role in traction control systems, which monitor wheel speed to prevent wheel slip or loss of traction. The system can adjust power delivery or apply brakes to specific wheels to maintain traction. ESC systems use wheel speed data from tone rings to help prevent skidding or loss of control by selectively applying brakes to specific wheels.

Tone rings work in conjunction with sensors, such as wheel speed sensors or magnetic sensors, which are positioned close to the rotating tone ring. As the tone ring's teeth pass by the sensor, they create fluctuations in the magnetic field, generating an electrical signal that is sent to the vehicle's ECM. The ECM uses this signal to calculate the wheel's speed and detect any variations in speed between wheels.

Tone rings can vary in design, including the number of teeth or notches they possess, their shape, and their mounting location. The spacing and pattern of the teeth or notches determine the frequency of the electrical signal generated, which is used to calculate the rotational speed.

Torque

Torque refers to the rotational force exerted by an engine on its crankshaft or other rotating components. Torque is a critical measurement in understanding the engine's ability to generate power and perform tasks such as moving heavy loads, accelerating, and overcoming resistance. Torque is a twisting or turning force applied to a shaft or axis of rotation. It is what causes an object to rotate around an axis.

Tour Bus

Tour buses are commercial vehicles designed specifically for providing guided tours to groups of passengers. These buses are equipped with amenities and features that enhance the travel experience, making them suitable for sightseeing, city tours, cultural explorations, and other guided trips.

Tour buses are designed with comfortable seating arrangements to ensure a pleasant and enjoyable ride for passengers during the tour (Figure T.2). These buses feature large windows to offer passengers unobstructed views of landmarks, attractions, and scenic spots. Many tour buses provide onboard commentary or audio guides that offer information about the sites being visited, enhancing the educational aspect of the tour. Depending on the type of tour and operator, tour buses may have amenities such as air conditioning, restrooms, and entertainment systems. Some tour buses have designated storage areas for passengers' personal items and luggage.

FIGURE T.2 An example of a tour bus.

Rob Wilson/Shutterstock.com.

Tow Dolly

A tow dolly refers to a type of trailer or small platform designed to assist in towing a vehicle behind another vehicle. Tow dollies are used to transport a front-wheel drive vehicle or a vehicle that cannot be driven on its own due to mechanical issues. They provide a means of towing without having to use a full-size trailer.

Tow dollies have a two-wheel configuration, where the towed vehicle's front wheels are placed on the dolly and secured. Tow dollies are equipped with a mechanism to ensure the towed vehicle's front follows the towing vehicle, preventing the towed vehicle from errant movement during towing. Some tow dollies have steerable axles, allowing for easier maneuverability during towing. Tow dollies are generally lighter and more compact than full-size trailers, making them easier to store and transport when not in use. Tow dollies often have a loading ramp that allows the towed vehicle's front wheels to be driven or rolled onto the dolly. Tow dollies may include safety chains, straps, and other features to ensure the secure attachment of the towed vehicle.

Tow Truck

A tow truck is a specialized motor vehicle equipped with mechanisms and equipment designed to transport and recover disabled, immobile, or improperly parked vehicles. Tow trucks are crucial in roadside assistance, recovery, and ensuring vehicles' safe and efficient removal from various situations.

Tow trucks typically exhibit specific characteristics. Tow trucks are equipped with various types of towing equipment, including winches, cables, hooks, and hydraulic systems, which enable them to lift, secure, and transport vehicles. Tow trucks come in different body configurations, such as flatbeds (also known as rollback trucks) and traditional wheel-lift or hook-and-chain setups. Many tow trucks are equipped with powerful winches that can pull and load vehicles onto the tow truck's bed or platform. Hydraulic controls allow operators to lift, tilt, and manipulate the towing apparatus to secure vehicles safely.

Tow trucks often have storage compartments for tools, safety equipment, and towing accessories. Tow trucks may have flashing lights, reflective markings, and other warning signals to alert other road users to their presence. Tow trucks are designed with safety features to prevent accidents during the towing process, including stabilizers and safety chains.

Towing Capacity

Towing capacity refers to the maximum weight a vehicle can safely tow. It represents the heaviest load a vehicle, such as a truck or SUV, can pull while maintaining proper control, stability, and safety on the road.

Towing capacity is a critical specification for vehicles used for towing trailers, boats, RVs, and other heavy loads. It ensures that the vehicle can handle the additional weight without compromising its performance, braking capabilities, or structural integrity. Exceeding the towing capacity can lead to dangerous situations, such as reduced braking efficiency, instability, or damage to the vehicle.

Several factors influence a vehicle's towing capacity. Vehicles with more powerful engines and higher torque output are generally better equipped to handle heavier loads. Vehicles with transmissions designed for towing often have specialized gear ratios that help manage the added weight. Vehicles intended for towing typically have stronger chassis and suspension systems to support the additional load.

Towing generates heat, so vehicles designed for towing often have enhanced cooling systems to prevent overheating. Towing adds stress to the braking system, so vehicles with higher towing capacities often have larger brakes for better stopping power. Vehicles with higher axle ratios can transmit more torque to the wheels, aiding in towing heavier loads. Vehicles with reinforced frames and structural components are better equipped to handle the stresses of towing.

Towing capacity is typically specified in pounds (lbs) or kilograms (kg) and is provided by the vehicle manufacturer. It is crucial to consider both the vehicle's towing capacity and the specific requirements of the load being towed, including trailer weight and the weight of any cargo carried on the trailer.

TPMS (Tire Pressure Monitoring System)

A TPMS refers to an electronic system designed to monitor and display the air pressure of a vehicle's tires in real time. The primary purpose of a TPMS is to ensure that the tires are properly inflated, promoting safety, improving fuel efficiency, and extending tire life [2].

TPMS continuously monitors the air pressure of each tire and alerts the driver if the pressure falls below a certain threshold, indicating underinflation. Properly inflated tires contribute to better traction, braking performance, and vehicle stability. TPMS helps prevent accidents caused by underinflated tires. Underinflated tires create higher rolling resistance, reducing fuel efficiency. TPMS helps maintain optimal tire pressure for better fuel economy. Underinflated or overinflated tires wear unevenly and can lead to premature tire failure. TPMS helps extend tire life by promoting even wear. Some advanced TPMS also monitor tire temperature, providing insights into potential issues caused by overheating [3].

There are two main types of TPMS: the direct TPMS—this system uses pressure sensors mounted inside each tire to directly measure tire pressure and transmit data to the vehicle's ECM; and the indirect TPMS—this system uses the vehicle's existing sensors, such as wheel speed sensors from the ABS, to indirectly estimate tire pressure based on wheel speed and other factors.

A TPMS typically consists of numerous components. In the case of direct TPMS, pressure sensors are mounted inside each tire to measure pressure and transmit data wirelessly. The receiver control module receives data from the pressure sensors and processes the information to determine whether tire pressure is within the recommended range. TPMS displays a warning on the vehicle's dashboard if any tire's pressure falls below the recommended threshold. The TPMS alerts the driver through visual and audible warnings when tire pressure is significantly low.

In many regions, including the US and the European Union, TPMS is mandatory for all new vehicles due to its safety benefits and impact on fuel efficiency.

Tractor

An agricultural tractor refers to a specialized motorized vehicle designed primarily for use in agricultural operations (Figure T.3). These tractors are versatile machines that perform various tasks on farms and agricultural fields, such as plowing, tilling, planting, cultivating, and hauling.

FIGURE T.3 Agricultural tractors pull a variety of attachments to work various fields—this one is for cereal grains.

MAVV/Shutterstock.com.

Agricultural tractors have powerful engines in various horsepower ranges to suit different farming tasks. Tractors often have large, rugged tires or tracks to provide optimal traction on different types of terrain. Many agricultural tractors feature a three-point hitch system that allows the attachment of implements and tools, such as plows, cultivators, and seeders. Tractors have a PTO shaft that provides mechanical power to operate machinery like mowers, balers, and hay cutters. Some modern agricultural tractors have enclosed cabs with climate control, while others have an open design.

Agricultural tractors are used for various tasks in farming and agriculture. Tractors are used for plowing, tilling, harrowing, and cultivating fields to prepare them for planting. Tractors are equipped with seeders or planters to sow seeds and plant crops. They are used for tasks like applying fertilizers, pesticides, and herbicides to crops. Tractors can be used in conjunction with harvesting machinery for tasks such as cutting and collecting crops. Tractors are employed for hauling equipment, feeding, and harvesting crops on the farm.

Tractor-Mounted Sprayer

A tractor-mounted sprayer is a specialized commercial vehicle attachment used in agriculture, horticulture, and landscaping to precisely apply liquid pesticides, herbicides, fertilizers, and other chemicals to crops, orchards, vineyards, or other vegetation. These sprayers are typically attached to a tractor, which provides the necessary power and mobility for effective and efficient spraying operations.

A tractor-mounted sprayer is designed to be easily attached to the rear or front of a tractor using a three-point hitch system or other standardized attachment mechanisms. Tractor-mounted sprayers come in various tank sizes to accommodate different spraying needs, ranging from small-scale horticulture to large-scale agriculture. These sprayers are equipped with a series of nozzles and adjustable booms (arms) that distribute the spray evenly over the target area. The number and configuration of nozzles and booms can vary depending on the specific application.

Many tractor-mounted sprayers use hydraulic systems to control the spraying functions, including nozzle adjustment, boom height, and flow rate. This allows operators to make precise adjustments while on the go. A pump within the sprayer pressurizes the liquid in the tank and delivers it to the nozzles. An agitation system ensures that the chemical mixture remains well-mixed and uniform. Tractor-mounted sprayers often feature GPS technology and variable rate control systems, enabling precise application based on field mapping and crop-specific requirements. This minimizes the overuse of chemicals and reduces environmental impact. Sprayers can have different boom configurations, including foldable, telescopic, or multi-section booms, to adjust the spray width and pattern according to the crop or terrain.

The height of the spraying boom can be adjusted to maintain the optimal distance from the target, ensuring thorough coverage while reducing drift. Operators typically have access to a control panel within the tractor cabin, allowing them to monitor and adjust spraying parameters, such as pressure, flow rate, and nozzle settings. Safety considerations are crucial in sprayer design. Features like boom breakaway systems and tank rinsing capabilities help prevent accidents and chemical residue buildup. Operators must adhere to environmental regulations when handling and applying chemicals using tractor-mounted sprayers to minimize contamination and runoff.

Tractor Protection Valve (Air Brakes)

This air brake air control valve protects the tractor air supply in the event of a trailer breakaway or total air loss. Two airlines connect the tractor brakes to the trailer air circuit. The trailer path supplies the trailer with air for braking and the other pneumatic systems like its suspension. The trailer line is the service brake signal line. The trailer lines are connected to the tractor protection valve. When the trailer supply dash valve is in the off position, the tractor protection valve is off. When the drivers make a service brake application, no air will exit the tractor protection valve service signal line to the trailer. When the trailer air supply valve is open, this air will open the tractor protection valve and provide the trailer with air. When a service brake application is made, the service brake signal air is transmitted through the tractor protection valve to use its service brakes.

For more information on this, see **J1410_202012** [4].

Tractor–Trailer

A tractor–trailer refers to a combination of two separate components: a tractor (also known as a truck or prime mover) and a trailer. This combination is widely used for transporting goods over long distances and is a common sight on highways and roads.

The tractor, also known as the truck or prime mover, is the front part of the combination. It includes the engine, driver's compartment (cab), and various systems necessary for operation. Tractors have a powerful engine that provides the torque required to haul heavy loads. The driver's compartment contains controls, seating for the driver, and sometimes additional passengers. Tractors can vary in size, power, and configuration and are responsible for pulling and propelling the entire combination.

The trailer is the rear portion of the combination and is designed to carry cargo. Trailers come in various types and sizes to accommodate different types of goods and transportation needs. Common trailers include flatbeds, dry vans, refrigerated trailers, and tanker trailers. Trailers can be loaded with a wide range of products, from general freight to specialized cargo, and their design is tailored to the specific requirements of the cargo being transported.

The tractor–trailer combination connects the trailer to the tractor using a coupling system, such as a fifth-wheel hitch. The tractor provides the power and control necessary to move and steer the combination while the trailer carries the cargo. The articulated nature of the combination allows for flexibility in turning and maneuvering. Tractor–trailers are commonly used for long-haul transportation, delivering goods between distribution centers, warehouses, and retail locations.

Trailer

A trailer refers to a non-motorized vehicle designed to be towed by a powered vehicle, such as a truck or a car. Trailers transport goods, materials, equipment, or other cargo in various sizes, types, and configurations to suit different transportation needs.

There are different types of trailers, including flatbed trailers, enclosed trailers, utility trailers, refrigerated trailers, and specialized trailers designed for specific cargo. Trailers have a specific payload capacity, indicating the maximum cargo weight they can safely carry. A hitch or coupling mechanism attaches the trailer to the towing vehicle. Trailers have axles and wheels to support the cargo's weight and facilitate movement.

Trailer Brake

A trailer brake refers to a braking system installed on a trailer that works in conjunction with the towing vehicle's brakes. Trailer brakes are essential for enhancing the safety, stability, and control of the towing vehicle and trailer combination during braking maneuvers.

The primary function of a trailer brake is to assist in slowing down or stopping the trailer when the towing vehicle's brakes are applied. When the driver of the towing vehicle applies the brakes, the trailer brake system responds by using braking force to the trailer's wheels, reducing the risk of skidding, swaying, or jackknifing.

There are different types of trailer brake systems, each with its own features and benefits. Electric trailer brakes use an electrical connection between the towing vehicle and the trailer to activate the trailer's braking system. When the driver applies the towing vehicle's brakes, the trailer's electric brakes engage, contributing to a controlled and synchronized braking process. Surge brakes are commonly used on trailers with smaller loads. They use the trailer's own momentum to engage the brakes when the towing vehicle slows down, creating hydraulic pressure to activate the trailer's braking system. Some heavy-duty trailers, especially those used in commercial applications like semi-trailers, use air brakes similar to those found on large trucks. These systems use compressed air to activate the trailer's brakes.

Trailer Hand Brake (Air Brakes)

The trailer control valve is employed to actuate the trailer service brakes on a trailer independently of the tractor service brakes. This valve is additionally known by the terms trolley valve, broker brake, and spike. The source of air supplied to the trailer control valve is typically the secondary circuit. The pressure of air is modulated proportionally with control valve travel. Some drivers are known to use this valve to incorrectly stop a semi using just the trailer brakes.

For more information on this, see **J1410_202012** [4].

Trailer Hitch and Connections

A trailer hitch and its associated connections refer to the mechanical components and systems used to attach a trailer to a towing vehicle, such as a truck or van. These components enable safe and secure trailer coupling to the towing vehicle, allowing the combined unit to be maneuvered and controlled as a single entity.

A trailer hitch is a device mounted to the rear of a towing vehicle that provides a connection point for attaching a trailer. It comprises various components designed to link the towing vehicle and trailer securely. Trailer hitches come in different types and classes, each with its capacity and design to accommodate varying loads and trailer types.

There are a number of common types of trailer hitches. The receiver hitch type features a receiver tube mounted to the vehicle's frame. It allows different hitch accessories, such as ball mounts or hitch-mounted bike racks, to be inserted into the receiver. The fifth-wheel hitch is commonly used with larger trailers, and fifth-wheel hitches are mounted in a pickup truck's bed and designed for heavier loads, providing increased stability and control. Similar to a fifth-wheel hitch, a gooseneck hitch is mounted in the bed of a truck. It uses a ball-and-coupler system for towing, often used with heavy trailers like livestock or equipment trailers.

Trailer Tracking

Tracking the movements and position of an articulated tractor–trailer unit through the use of a location unit fitted to the trailer, and a method of returning the position data via a mobile communication network or geostationary satellite communications for use through either a personal computer or cloud-based software. Freight containers can also be tracked by GPS using a similar approach to that used for trailer tracking like a battery-powered GPS device communicating its position using a cell phone. This use provides increased security and the possibility to reschedule the container transport movements based on accurate information about their location.

Transit Bus

Transit buses refer to large motor vehicles designed primarily for the public transportation of passengers within urban or suburban areas. Transit buses are a crucial component of public transit systems. They provide a reliable and efficient means of moving many passengers from one location to another within a city or its surrounding regions.

Transit buses are commonly used in urban and suburban areas. Transit buses serve as the backbone of public transit systems, providing an affordable and efficient mode of transportation for residents and commuters. Transit buses connect residential areas to commercial centers, allowing commuters to travel to work, schools, and other destinations. In some locations, transit buses offer guided tours of city attractions and landmarks. Transit buses are often used to transport people to and from special events, concerts, and sporting venues. Using transit buses can reduce traffic congestion and emissions, promoting sustainable urban transportation.

Transmission

Transmission refers to the mechanical component of a motor vehicle that transfers power from the engine to the wheels, enabling the vehicle to change speeds and move forward or backward. The transmission system is critical in controlling the vehicle's speed, power distribution, and efficiency by selecting different gear ratios.

The primary function of a transmission is to convert the rotational power generated by the engine into the appropriate torque and speed required for various driving conditions. The transmission achieves this by using different gear ratios that allow the engine to operate efficiently across various speeds.

There are two main types of transmissions used in commercial vehicles, manual and automatic. Manual transmissions are also known as a stick shift or manual gearbox, and this type of transmission requires the driver to engage manually and disengage gears using a clutch pedal and gear shifter. Manual transmissions offer more control over gear selection and are commonly used in commercial vehicles.

An automatic transmission shifts gears automatically without requiring driver intervention. It uses a hydraulic system, electronic controls, and a torque converter to manage gear changes based on vehicle speed and load. Automatic transmissions are popular for their convenience and ease of use.

The choice between manual and automatic transmissions depends on the vehicle's intended use and the industry's specific needs. Commercial vehicles, such as trucks, buses, vans, and delivery vehicles, may have varying transmission requirements based on load capacity, terrain, and driver preference.

Modern commercial vehicles often incorporate advanced transmission control systems. ECUs manage the shifting process in automatic transmissions and can optimize gear changes for fuel efficiency and performance.

AMTs combine elements of both manual and automatic transmissions, using computer-controlled actuators to shift gears without a clutch pedal. Continuous variable transmissions (CVTs), used in agricultural equipment, provide a constant range of gear ratios, leading to smoother acceleration and improved fuel economy.

Transmission Brake

A transmission brake in the context of commercial vehicles refers to a type of braking system that is integrated into the vehicle's transmission. This system is designed to provide additional braking force and assist in controlling the vehicle's speed, especially during descents or when slowing down. The transmission brake works by engaging the transmission to restrict the movement of the drivetrain, helping to slow down the vehicle without relying solely on the wheel brakes.

The use of a transmission brake can contribute to improved vehicle control, enhanced safety, and reduced wear on the wheel brakes. It is particularly useful in situations where prolonged braking is required, such as when navigating steep slopes or long descents. Additionally, the transmission brake can be a valuable component in combination with other braking systems to achieve optimal braking performance in commercial vehicles.

For more information, see **J2688** [5].

Truck

A truck is a motor vehicle designed primarily to transport goods, materials, and cargo. Trucks are essential assets in various industries, enabling the efficient movement of goods over short and long distances. They come in multiple sizes, configurations, and capacities to accommodate loads and operational needs.

Trucks typically exhibit consistent characteristics. Trucks are characterized by a cargo bed or loading area behind the driver's compartment. This space is designed to carry goods, equipment, materials, or other loads. Trucks have varying payload capacities, indicating the maximum weight of cargo they can carry, including the vehicle's weight. Trucks may have single or multiple axles, and their configuration affects load distribution, stability, and maneuverability.

The driver's compartment, or cab, can come in various styles, including standard, extended, and crew cabs, depending on the number of occupants and comfort requirements. Trucks can have the engine placed in the front (conventional) or COE configuration, which impacts visibility, space utilization, and handling. There are specialized truck types, such as flatbed trucks (for carrying large or heavy loads), dump trucks (for transporting loose materials), and tanker trucks (for liquids or gases). Some trucks are equipped with towing capabilities to pull trailers or other heavy loads.

Truck Frame

A truck frame consists of two large pieces of longitudinal angle or channel, most commonly steel and aluminum for light weight applications or a combination of both, with a number of crossmember steel rails that are welded, riveted, or bolted together to form the vehicle component support system, such as axles and suspension systems. On a tractor, the trailer weight is added to the fifth-wheel assembly. On a straight-body truck, the GVW is distributed across the total length and width of the frame. Just how much force the frame absorbs depends on where the front and rear axle(s) are located and how the load is distributed.

Turbocharger

A turbocharger is a mechanical device used to increase an ICE's efficiency and power output by forcing more air into the engine's combustion chambers. This additional air allows for more efficient fuel combustion, improving engine performance.

A turbocharger operates by utilizing the exhaust gases generated by the engine to spin a turbine, which is connected to a compressor. The compressor draws in and compresses outside air, which is then forced into the engine's cylinders at a higher pressure and density. This process, known as forced induction, allows the engine to burn more fuel efficiently, producing more power and torque without needing a larger engine displacement.

Two-Gear Reduction Axle

A two-gear reduction axle is a final drive rear axle that has two separate output ratios available from the same differential assembly. Much like a 4WD system found in an SUV, the driver is able to switch between each output mode using a switch in the truck.

Two-Pedal System

Two-pedal system is an automated manual truck transmission system that does not use a clutch pedal. The two pedal system refers to the operation of a vehicle with accelerator and brake pedals, which is typically associated with an automatic transmission, or automated manual transmission.

Two-Stroke Engine

A two-stroke engine also goes through intake compression, power, and exhaust actions to finish one operating cycle. However, the intake and compression actions are combined in one stroke, and the power and exhaust actions are combined within the other stroke. Two-stroke engines are utilized in lawnmowers, heavy trucks, construction equipment, and ships.

When compared to the four-stroke cycle, the two-stroke cycle completes two strokes or actions in one turn (revolution or 360 degrees) of the engine (i.e., compression and power). The following items apply to the two-stroke cycle:

- The intake valve has been eliminated and replaced by a row of ports around the cylinder liner.
- A root-type blower is placed on the engine to blow air into the cylinder whenever the piston uncovers the ports. These ports have an angular inlet called the Hesselman helical loop to swirl the air creating turbulence with a resultant improvement in the air–fuel mixing. The exhaust valve or valves open to allow the blower to push the exhaust gases out of the cylinder. The blower blows intake air alone into the cylinder when the ports are uncovered. The piston and the exhaust valve close and then cover the ports. The compression of the air takes place as the piston approaches the top dead center (TDC). The injector is timed to inject fuel, finely atomized, into the hot compressed air in the cylinder. The rapid expansion of the burning gases forces the piston down toward the bottom dead center (BDC), and the exhaust valve opens. The piston uncovers the inlet ports, and the cylinders are swept clean with scavenging air.

REFERENCES

1. Natteri, A., Surendran, J., Paul Raj Bob, B., Pushparaj, K. et al., "A Telematics Enabled Analytics Approach for Determining Tractor Usage," *SAE Int. J. Adv. & Curr. Prac. in Mobility* 4, no. 3 (2022): 734-741, doi:https://doi.org/10.4271/2021-26-0089.

2. Society of Automotive Engineers, "J2848 Tire Pressure Systems - Management (CTIS) Type for Medium- and Heavy-Duty Highway Vehicles," SAE Publishing, Warrendale, 2021.

3. US Department of Transportation, "Tire Pressure Monitoring System FMVSS No. 138," US Department of Transportation, Washington, DC, 2001.

4. Society of Automotive Engineers, "Standard J1410_202012: Air Brake Valve - Performance Requirements," SAE Publishing, Warrendale, 2020.

5. Society of Automotive Engineers, "J2688 Parking Brake Control Identification - Vehicles with Hydraulic Brake Systems and Automatic Transmissions," SAE Publishing, Warrendale, 2020.

U

Ultra-Capacitor Module

An ultra-capacitor module is a control module that supplements the cranking current for a 24-V battery system on a heavy-duty vehicle, which reduces the starting time. It is also known as a supercapacitor.

UltraShift Truck Transmission

The UltraShift electrically controlled manual transmission is fully automatic. Although they use a clutch, a no-clutch pedal is required; hence, they are referred to as two-pedal units because they use a clutch that is electro-pneumatically actuated to link the clutch electronically to the management computer and pushbutton or tower-type shift controls. Both the semi-automated and fully automated transmission types must be mated to an electronically managed engine using the **SAE J-1939** Controller Area Network (CAN) standard between the shift control and the engine management computer.

The Eaton fully automatic UltraShift transmission evolved from the semi-automated AutoSelect and AutoShift models. UltraShift models do not require a clutch pedal because they use an electro-pneumatically actuated clutch to link the clutch electronically to the transmission management computer. All semi-automated and fully automated or tower-type models offer shift-by-wire controls. The ten-speed UltraShift model uses an Eaton Fuller 15-1/2-in. two-plate datalink mechanical clutch that is computer-controlled and a high-capacity inertia brake to speed automated upshifts. The inertia brake also serves as a clutch brake. The UltraShift models are also designed with advanced generation II electronics. The UltraShift can be used in fully automatic mode or with manual intervention, according to the driver's preference.

The ten-speed UltraShift model provides automatic shifting as operating and load conditions allow, plus float shifting between gears to extend clutch life. The UltraShift transmission uses a three-position pushbutton selector, which offers a manual mode to hold the current gear and permits driver-selected upshifts or downshifts. The low mode feature can be used in any gear to prevent the transmission from upshifting and initiate downshifts to optimize engine braking performance. The low mode can also be used to select the first gear at a standstill, which will hold in low until drive mode is selected. The low mode also offers maximum engine braking.

Unitized Hub

To reduce the service requirements of wheel hubs, a unitized wheel end is a pre-packaged system in which the bearings, seals, and lubricant are pre-installed into the hub. Unitized hubs can be filled with synthetic grease and sealed to avoid lubrication intervals, bearing adjustment, and seal replacement, thereby eliminating hours of typical wheel hub maintenance. The bearings can be either integrated into the hub or pressed in as a separate cartridge. The bearing adjustment is set during manufacturing, and the hub nut retains the assembled hub on the axle spindle. This hub cannot be serviced in the field but must be returned to the manufacturer for rebuilding or exchange.

Unitized Seal

One-piece unitized seals are most commonly found in today's heavy-duty truck applications. This design incorporates a wear sleeve and seal into a single unit, simplifying installation. The seal is pressed into the hub and rotates with the hub.

Utility Tractor

A utility tractor is a versatile, rugged commercial vehicle designed primarily for agricultural and landscaping tasks. These tractors are characterized by their adaptability and ability to handle a wide range of attachments and implements, making them valuable tools for various farming, construction, and grounds maintenance applications.

Utility tractors come in various sizes and power ratings, typically measured in HP, to accommodate different tasks and workloads. Smaller models are suitable for light-duty work, while larger ones can handle heavier applications. The defining feature of utility tractors is their adaptability. They are designed to accept a variety of attachments and implements, such as front-end loaders, backhoes, plows, mowers, and tillers, allowing operators to perform a wide range of tasks.

Many utility tractors are equipped with 4WD to improve traction and stability, making them suitable for work in challenging terrains and adverse weather conditions. Utility tractors typically have a PTO that transfers power to attached implements, enabling functions like mowing, tilling, and powering auxiliary equipment. A three-point hitch system at the tractor's rear provides a standardized way to attach and control various implements, ensuring compatibility with a wide range of tools.

Modern utility tractors often feature a comfortable and ergonomic operator station with climate control, adjustable seating, and intuitive controls for ease of use. Hydraulic systems on utility tractor power implements like loaders allow operators to raise, lower, and control attachments precisely. Some utility tractors offer enclosed cabs for added operator comfort and protection from the elements. Utility tractors can have various tire types and treads to suit different terrains and applications.

Given their demanding tasks and environments, regular maintenance is crucial to keeping a utility tractor in optimal working condition. Utility tractors are used in a wide range of applications, including farming (e.g., plowing, planting, and harvesting), landscaping (e.g., mowing and digging), construction (e.g., digging and material handling), and property maintenance (e.g., snow removal and brush clearing).

The choice of attachments and implements greatly expands the utility tractor's capabilities. Operators can switch between tasks by attaching and detaching the necessary equipment. Utility tractors may run on diesel fuel or, in some cases, alternative fuels, providing flexibility for operators with specific fuel preferences or environmental considerations.

Utility Trailer

Utility trailers refer to non-motorized, wheeled vehicles designed to be towed behind a motorized vehicle. Utility trailers transport various types of cargo, equipment, and materials, offering additional storage and hauling capacity to complement a towing vehicle.

Utility Truck

A utility truck refers to a specialized type of motor vehicle designed for tasks that require the transportation of equipment, tools, and materials needed for maintenance, repair, construction, and other utility-related activities. Utility trucks are equipped with various features, compartments, and systems to enable efficient and organized work in various industries.

Utility Vehicle

A utility vehicle refers to a type of motor vehicle that is designed and equipped to perform a wide range of tasks and functions, particularly in professional and industrial settings. Utility vehicles are known for their versatility, durability, and adaptability, making them essential tools for various commercial applications.

Utility vehicles typically exhibit some common characteristics. Utility vehicles are designed to handle diverse tasks and applications. They can have various attachments, tools, or equipment to fulfill specific functions. Utility vehicles are built to withstand rugged conditions and heavy use, making them suitable for demanding work environments. Many utility vehicles are designed for off-road operation, enabling them to navigate rough terrains and challenging landscapes.

Utility vehicles often have a substantial payload capacity, allowing them to transport equipment, materials, and tools. Some utility vehicles are equipped with towing capabilities, enabling them to tow trailers, implements, or other heavy loads. Many utility vehicles are compact in size, making them maneuverable in tight spaces, such as construction sites or warehouses. Utility vehicles can have open or closed cabs, depending on the intended use and the need for operator protection.

Utility vehicles find applications across a wide range of industries and sectors such as construction, agriculture, landscaping, facility maintenance, warehousing and logistics, and recreation.

V

Vacuum Truck

A vacuum truck is a specialized vehicle equipped with a powerful vacuum system designed to suction and transport various liquids, sludges, solids, and debris from multiple environments. Vacuum trucks are commonly used for waste management, environmental cleanup, industrial applications, and maintaining sanitary systems.

Vacuum trucks are equipped with a vacuum pump or blower that creates a strong suction force, allowing them to remove materials from tanks, containers, sewers, and other spaces. These trucks have storage tanks to hold the collected materials. Tanks are often made of durable materials to handle various substances. Vacuum trucks use flexible suction hoses to reach confined spaces and extract materials. Some vacuum trucks have a hydraulic dumping mechanism to discharge the collected materials at designated disposal sites. Some models are equipped with a high-pressure water system for cleaning purposes, commonly used in combination with the vacuum system.

Vacuum trucks have a range of applications. Vacuum trucks collect and transport liquid waste, sludges, and solids from septic tanks, grease traps, and wastewater treatment facilities. They are employed in industrial settings to clean tanks, pits, and other spaces that may contain hazardous materials.

Vacuum trucks are crucial in cleaning up spills, hazardous waste, and contaminated sites. Municipalities use vacuum trucks to clean and maintain sewer lines, storm drains, and catch basins. Vacuum trucks assist in cleaning construction sites by removing excess water, debris, and waste materials. Vacuum trucks are used to clean and maintain storage tanks, sumps, and other equipment in the oil and gas industry.

Van

A van refers to a type of motor vehicle primarily designed to transport goods, equipment, or passengers. Vans typically have a closed-body structure with a rear cargo area fully enclosed by walls and a roof. Vans are sometimes widely used in various industries to transport goods, tools, equipment, and passengers.

Vans share several characteristics that distinguish them from other types of vehicles. Vans feature a fully enclosed cargo area, protecting the contents from the elements and providing security. Vans come in various sizes, from compact city vans to larger high-capacity vans, allowing businesses to choose a model that suits their needs. Vans are designed to carry various payloads, with their capacity ranging from light duty to heavy duty, depending on the van's size and configuration. Vans can be configured with different door arrangements, including sliding side doors and rear doors that swing open or lift up, facilitating efficient loading and unloading. Many vans offer features to help organize cargo, such as tie-down points, shelving, and partitions to separate the cargo area from the driver's compartment. Some vans are designed with additional seating for passengers, making them suitable for group transportation, such as shuttle or small-scale public transit.

Vans find applications in a wide range of industries and services. Vans are commonly used to deliver packages, parcels, and goods to customers or between locations. Tradespeople and contractors use vans to transport tools, equipment, and materials to job sites. Vans are used to transport merchandise for mobile businesses, vendors, and event marketing. Service providers like plumbers, electricians, and technicians often use vans to carry tools and supplies for on-site work. Some vans are adapted for passenger transportation, serving as shuttles, small buses, or people carriers. Camper vans and RVs have been converted for living and traveling purposes.

Variable Rate Planting

Variable rate planting is a technique that adjusts planting rates based on GPS data to optimize seed placement and spacing according to soil conditions and yield potential [1].

A commercial variable rate planting vehicle is a specialized agricultural machine equipped with technology and software that precisely adjusts crops' planting rates as it moves across a field. This adjustment is made in real time based on data gathered from various sources, such as soil maps, yield history, or sensor data. The primary goal of variable rate planting is to optimize seed placement and spacing to match the specific conditions and requirements of different areas within a field.

There are a few key components of a commercial variable rate planting vehicle. These vehicles often use GPS or similar satellite-based systems to map the field and accurately determine the machine's position. A variable rate control system processes data from various sources to adjust the planting rate as the vehicle moves through the field. Precision seed meters are used to control the flow of seeds and adjust planting rates according to the prescribed rates.

Specialized software helps interpret and integrate data from soil tests, yield maps, and other sources to make real-time planting rate decisions. Operators can interact with the control system to monitor and adjust planting rates as needed.

Commercial variable rate planting vehicles are a part of precision agriculture practices and offer several benefits, including optimized use of seeds, increased crop uniformity, improved yield potential, and resource efficiency. These vehicles are valuable tools for modern farming operations seeking to maximize productivity while minimizing input costs.

Vegetable Transplanter

A commercial vegetable transplanter is a specialized agricultural vehicle or implement equipped with mechanisms designed to plant young vegetable seedlings into prepared fields or growing beds. These machines are commonly used in large-scale vegetable farming operations to streamline the transplanting process, reduce labor requirements, and ensure the seedlings' proper spacing and planting depth. Vegetable transplanters can handle a variety of vegetable crops, including tomatoes, peppers, cabbage, and lettuce.

A commercial vegetable transplanter typically consists of a conveyor system or planting mechanism that moves trays or individual seedlings from a hopper to the planting location. Depending on the design, these transplanters may use vacuum or mechanical mechanisms to carefully place seedlings in the soil at the desired intervals and depths. Some transplanters can also apply water or fertilizer as part of the planting process to enhance seedling establishment.

Vehicle Classification

Vehicle classification refers to the categorization of different types of commercial vehicles based on specific criteria such as size, weight, purpose, and design. This classification system helps organize and standardize commercial vehicles for regulatory, administrative, and operational goals.

Vehicle classification serves several purposes. Different classes of commercial vehicles are subject to specific regulations, standards, and requirements based on their classification. This helps ensure safety, efficiency, and environmental compliance. Vehicle classification determines the license and registration type needed for legally operating a commercial vehicle on the road. Urban planning, road maintenance, and infrastructure development are influenced by the types and sizes of vehicles that use the roads. Classification data help authorities plan and design roadways accordingly. Vehicle classification assists in traffic management, toll collection, and congestion reduction by categorizing vehicles for various purposes. Collecting data on vehicle classifications aids in transportation research, statistics, and policy development.

Commercial vehicles are classified based on a few factors. Classification is based on the maximum weight (GVWR) a vehicle is rated to carry, including its weight and payload. Vehicles are categorized by their body styles, such as trucks, vans, buses, and trailers. Classification can be based on the primary function of the vehicle, such as freight transport, passenger transport, or specialized purposes like construction or refrigeration. Vehicles with varying numbers of axles fall into different classes based on their weight distribution and load-carrying capacity.

The number of seats in passenger vehicles can influence classification. Some classification systems consider vehicles' emission standards, particularly in relation to environmental regulations. The arrangement of axles, number of wheels, and specific features contribute to classification.

Common vehicle classes include light-duty, medium-duty, heavy-duty, buses, vans, and various categories of trailers.

Vehicle Configuration

Vehicle configuration refers to the specific arrangement, design, and setup of a commercial vehicle's components, features, and specifications. It encompasses various factors that define how the vehicle is built, including its body type, engine options, transmission, axle configuration, payload capacity, seating arrangement, and other customization choices.

Determining the optimal vehicle configuration is crucial for meeting the specific needs of different industries and applications. Commercial vehicles serve various purposes, such as transportation, delivery, and construction. The chosen configuration directly affects the vehicle's functionality, performance, efficiency, safety, and suitability for its intended tasks. Commercial vehicles come in various configurations to cater to different requirements. Body type refers to the style of the vehicle's body, such as trucks, vans, buses, and trailers. Each body type is designed to fulfill specific functions and accommodate different types of cargo or passengers. Commercial vehicles may offer multiple engine options with varying power outputs and fuel efficiencies. The engine choice affects the vehicle's performance and fuel economy. The transmission type, whether manual or automatic, influences how the vehicle shifts gears and operates, impacting driver comfort and fuel efficiency. The number of axles and their arrangement (single, tandem, tri-axle, etc.) affect the vehicle's load-carrying capacity and maneuverability. A vehicle's maximum weight, including cargo and passengers, influences its suitability for specific tasks. Depending on the intended use, commercial vehicles may be equipped with specialized features such as lift gates, refrigeration units, and cranes.

Vehicle Maintenance System

This system helps track and manage vehicle maintenance schedules, inspections, repairs, and preventive maintenance tasks, ensuring vehicles are in safe and operational condition.

VIN (Vehicle Identification Number)

The VIN is a unique alphanumeric code assigned to each commercial vehicle. The VIN serves as a distinctive identifier for the vehicle and contains essential information about its specifications, manufacturing details, and history. It plays a crucial role in tracking and documenting vehicles throughout their lifecycle, from production to registration, maintenance, and eventual disposal.

A VIN is typically composed of 17 characters, which can include a combination of letters, numbers, and sometimes special characters. Each character in the VIN represents specific information about the vehicle, such as its country of origin, manufacturer, vehicle type, engine size, and model year. The VIN is organized into sections that provide essential data points for anyone seeking to decode its information.

The VIN is used for various purposes, including vehicle registration, insurance, vehicle history reports, recalls, warranty claims, and law enforcement identification. It helps prevent theft, fraud, and stolen vehicle sales by providing a standardized and globally recognized means of identifying individual vehicles.

To decode the VIN and extract its information, various online tools and databases can provide details about a vehicle's make, model, year, engine type, production plant, and more.

REFERENCE

1. Natteri, A., Surendran, J., Paul Raj Bob, B., Pushparaj, K. et al., "A Telematics Enabled Analytics Approach for Determining Tractor Usage," *SAE Int. J. Adv. & Curr. Prac. in Mobility* 4, no. 3 (2022): 734-741, doi:https://doi.org/10.4271/2021-26-0089.

W

Water Truck

A water truck is a specialized vehicle designed to transport and distribute water for various purposes, such as dust suppression, construction site watering, road maintenance, and firefighting. Water trucks are equipped with tanks, pumps, and spray systems to deliver water to designated areas effectively.

Water trucks are equipped with a storage tank that holds a large volume of water. Tanks are often made of materials suitable for carrying and storing water. Water trucks have a pump system that enables water to be drawn from the tank and delivered through hoses and spray nozzles. These trucks are equipped with spray systems, including nozzles or hoses for distributing water. Spray patterns can be adjusted for different applications.

Many water trucks have a pressurized system that allows water to be sprayed over longer distances and heights. Some water trucks are equipped with water cannons that can project water over significant distances, making them useful for firefighting and dust suppression.

Water trucks have a range of applications. Water trucks are used to control dust on construction sites, unpaved roads, mining operations, and other areas prone to dust generation. Water trucks are used to keep roads damp to prevent dust, control soil erosion, and improve road compaction. They provide water for construction activities such as concrete curing and soil compaction.

Some water trucks are equipped for firefighting, delivering water to remote areas where fire hydrants may not be available. Water trucks are used to water lawns, plants, and trees in landscaping projects. Water trucks can be used in agriculture for irrigation, crop dusting, and animal care.

Waypoints

GPS coordinates are used as reference points for guiding planting equipment along specific routes or patterns.

Well-Testing Truck

A well-testing truck is a specialized commercial vehicle used in the oil and gas industry to conduct well-testing operations. These trucks are equipped with various tools and equipment necessary to assess the performance and characteristics of oil and gas wells.

Well-testing trucks are equipped with instruments and sensors to collect data such as pressure, flow rates, temperature, and fluid composition from oil and gas wells. These data help operators evaluate the well's productivity and reservoir properties. Operators can also take samples of reservoir fluids (oil, gas, and water) for analysis to determine their quality and properties.

Well-testing trucks can control and regulate the pressure within the well to ensure safe and efficient testing operations. These trucks often have equipment to separate oil, gas, and water from the well stream for individual analysis and measurement.

Accurate flow measurement devices are installed to quantify the oil and gas production rate from the well. Data collected during the testing process are recorded and analyzed to assess the well's potential, identify production issues, and make decisions regarding well management.

Well-testing trucks are designed with safety in mind, often including emergency shut-off systems and containment measures to mitigate potential hazards. They are typically mounted on truck chassis for easy transportation to well sites, making them mobile and versatile for testing operations at different locations.

Wet Tank

Traditionally, a wet tank refers to a reservoir that stores and supplies compressed air to various pneumatic systems within the vehicle, particularly air brake systems. The wet tank is a vital component of the air system that helps capture moisture content as the air from the compressor cools off.

The wet tank stores compressed air generated by the vehicle's air compressor. This compressed air is utilized to actuate the vehicle's air brakes, common in heavy-duty commercial vehicles.

A typical wet tank assembly includes components such as drain valves to remove moisture and contaminants that can accumulate in the compressed air, pressure gauges to monitor the air pressure, safety valves to prevent over-pressurization, and connections for air supply lines.

In general a wet tank also refers to the storage tank for onboard liquids, the most common being hydraulic oil for hydraulic systems.

Wheel and Hub Assemblies

Wheel and hub assemblies are integral to a commercial vehicle's suspension and wheel systems. They consist of several parts working together to enable the vehicle's wheels to rotate smoothly while supporting its weight. A typical wheel and hub assembly includes the following components:

- The central component attaches to the axle.
- The wheel hub houses the wheel bearings and provides a mounting point for the wheel.
- Bearings allow the wheel hub to rotate smoothly around the axle.
- Bearings reduce friction and wear during wheel movement.

Inner and outer bearing races are rings that hold the wheel bearings in place and help distribute the load evenly. Seals are used to protect the wheel bearings from dirt, water, and contaminants, helping to prolong their lifespan and maintain proper lubrication. Nuts secure the wheel to the wheel hub assembly firmly in place. A protective cover shields the wheel hub and bearings from dust and debris.

The wheel and hub assemblies support the vehicle's weight, provide a smooth ride, and maintain proper alignment. Regular maintenance and inspection of these assemblies are essential to ensure safe operation and extend their lifespan.

Wheelbase

The wheelbase of a commercial vehicle refers to the longitudinal distance between the center points of the front and rear axles. It is the measurement from the center of the front-wheel axle to the center of the rear wheel axle. The wheelbase is a critical dimension in determining the positioning of the axles and the overall length of the vehicle's chassis. It plays a significant role in the vehicle's stability, maneuverability, and handling characteristics.

A longer wheelbase is often associated with greater stability at higher speeds and a smoother ride, making it suitable for transporting heavy loads. On the other hand, a shorter wheelbase enhances the vehicle's ability to navigate tight turns and maneuver in confined spaces. The wheelbase length choice depends on the commercial vehicle's intended application, whether it is for long-haul transport, urban delivery, or specialized use.

Windshield Wipers and Washers

Windshield wipers and washers are essential components in a commercial vehicle that help maintain visibility and safety by clearing the windshield of rain, snow, dirt, and other debris (Figure W.1).

FIGURE W.1 Target windshield viewing area—standard cab [1].

Classification	F Dimension (mm)	F Dimension (inches)	Area	Angle Up (degrees)	Angle Down (degrees)	Angle Left (degrees)	Angle Right (degrees)
Truck Cab Behind Engine (CBE) Cab Alongside Engine (CAE)	0-1020	0-40	A	10	5	18	56
			B	5	3	14	53
			C	5	1	10	15
	1020-1270	40-50	A	8	7	18	56
			B	3	5	14	53
			C	3	3	10	15
	1270-Up	50-Up	A	6	9	18	56
			B	1	7	14	53
			C	1	5	10	15
Truck Cab Over Engine (COE)	1020-Up	40-Up	A	7.5	22	22	62
			B	3	22	22	62
			C	1	16	22	15
Bus, Cab Behind Engine (CBE) School and Commercial	1270-1520	50-60	A	7.5	22	22	62
			B	3	22	22	62
			C	1	16	22	15
Bus, Cab Over Engine (COE) Forward Control, School and Commercial	1270-1520	50-60	A	7	14	18	65
			B	2	11	18	65
			C	1	11	18	25
Forward Control or Multipurpose	All	All	A	9	7	18	56
			B	4	4	14	53
			C	2	2	10	15
Light-Duty Utility Vehicle	All	All	A	7	5	16	49
			B	4	3	13	49
			C	4	2	8	13
Van	Open	Open	A	7	12	18	58
			B	2	11	15	56
			C	1	6	10	15

Refer to SAE J687 for nomenclature.
Angles are minimum.

Windshield wipers are mechanical devices mounted on the exterior of a commercial vehicle's windshield. They consist of metal or plastic arms with rubber blades attached to them. When activated by the driver, usually through a control stalk inside the vehicle, the wipers move back and forth across the windshield's surface. The rubber blades contact the windshield, sweeping away water, snow, and debris to give the driver a clear line of sight.

Windshield washers spray a cleaning fluid, often a mixture of water and washer fluid, onto the vehicle's windshield. These systems typically consist of a reservoir containing the cleaning fluid, a pump to pressurize the fluid, hoses to carry the fluid to the windshield, and spray nozzles on the vehicle's hood or wiper arms. The driver can activate the windshield washers to spray the cleaning fluid onto the windshield, helping to dissolve and remove stubborn dirt and grime that the wiper blades might not effectively clear.

Wire Harness

A wire harness, also known as a wiring harness or cable harness, refers to a structured assembly of electrical wires, cables, connectors, and other components bound together to transmit electrical power and signals within a commercial vehicle. The primary purpose of a wire harness is to provide a reliable and organized method for routing and connecting various electrical circuits and components throughout the vehicle's systems, such as its engine, transmission, lighting, sensors, and entertainment systems.

Wire harnesses are essential in modern commercial vehicles as they help streamline the installation process of complex electrical systems, reduce the risk of wire damage or short circuits, enhance the overall safety and reliability of the vehicle, and simplify maintenance and troubleshooting tasks.

For more information, see **J2191**, **J1292**, **J2030**, and **J2202** [2, 3, 4, 5].

Wireline Truck (Logging and Well Intervention)

A wireline truck refers to a specialized commercial vehicle used in the oil and gas industry for wireline logging and intervention operations. Wireline trucks are equipped with the necessary equipment to lower and retrieve wireline tools, sensors, and instruments into wellbores for various purposes, including data collection, measurements, and well interventions.

Wireline trucks are equipped with hoisting equipment, such as winches, that lower and retrieve wireline tools and instruments into and out of the wellbore. These trucks have control systems that allow operators to monitor and control the deployment and movement of wireline tools and sensors.

Wireline trucks carry various tools and sensors for well logging, measurements, and interventions. These tools can include logging tools, pressure and temperature sensors, and tools for well interventions. Wireline trucks have reels of cabling that connect to the wireline tools and instruments being lowered into the well.

Wireline trucks are primarily used for various operations in the oil and gas industry. Wireline trucks are used for well logging, which involves lowering instruments into the well to collect data about the subsurface formations and reservoir properties. Wireline trucks can be used to monitor the flow of fluids within the well and assess production zones. Wireline trucks support well interventions, such as perforating, plug setting, and wellbore cleanout operations.

WMS (Warehouse Management System)

While not exclusively for vehicles, WMS software manages warehouse operations, including moving goods, storage, and order fulfillment, contributing to overall supply chain efficiency.

REFERENCES

1. Society of Automotive Engineers, "J198 Windshield Wiper Systems - Trucks, Buses, and Multipurpose Vehicles," SAE Publishing, Warrendale, 2020.
2. Society of Automotive Engineers, "J2191 Recommended Practice for Identification of Standardized Truck and Tractor Electrical Circuits," SAE Publishing, Warrendale, 2023.
3. Society of Automotive Engineers, "J1292 Automobile and Motor Coach Wiring," SAE Publishing, Warrendale, 2016.
4. Society of Automotive Engineers, "J2030 Heavy-Duty Electrical Connector Performance Standard," SAE Publishing, Warrendale, 2015.
5. Society of Automotive Engineers, "J2202 Heavy-Duty Wiring Systems for On-Highway Trucks," SAE Publishing, Warrendale, 2024.

Y

Yield Mapping

Yield mapping is a technique that uses GPS technology to record and map crop yields across a field, helping farmers identify areas of high and low productivity [1].

A commercial yield mapping vehicle is an agricultural machine equipped with sensors, GPS technology, and data recording systems designed to measure and record crop yields in real time as it traverses a field or agricultural area. The primary purpose of a yield mapping vehicle is to generate precise and detailed yield maps, which represent the variation in crop productivity within the field.

The system contains many components. These vehicles use GPS or GNSS technology to accurately determine their location and create georeferenced yield maps. Yield mapping vehicles are equipped with sensors, often mounted on the harvesting or crop collection equipment, that measure the quantity of harvested crops (e.g., grain, fruit, or vegetables) in real time. Data collected by the sensors are recorded and stored, often linked to GPS coordinates, allowing for detailed yield maps.

An onboard computer processes and interprets the sensor data, calculates yields, and may display real-time yield information to the operator. Yield mapping systems can integrate with other data sources, such as soil maps, weather data, and crop management records, to provide a comprehensive view of factors affecting crop productivity.

Commercial yield mapping vehicles are essential tools in precision agriculture, as they enable farmers to identify areas of a field with varying yield potential. This information can be used for making data-driven decisions related to crop management, such as adjusting irrigation, fertilizer application rates, and other practices to optimize yield and resource use efficiency. Yield maps also serve as valuable historical records for future planning and analysis.

REFERENCE

1. Natteri, A., Surendran, J., Paul Raj Bob, B., Pushparaj, K. et al., "A Telematics Enabled Analytics Approach for Determining Tractor Usage," *SAE Int. J. Adv. & Curr. Prac. in Mobility* 4, no. 3 (2022): 734-741, doi:https://doi.org/10.4271/2021-26-0089.

Index

About the Authors

Courtesy of Jon M. Quigley.

Jon M. Quigley PMP (204278) CTFL is a principal and founding member of Value Transformation, a product development (from idea to product retirement) and cost improvement organization established in 2009. Jon has an engineering degree from the University of North Carolina at Charlotte, two master's degrees from the City University of Seattle, and two globally recognized certifications. Jon has over 30 years of product development and manufacturing experience, ranging from embedded hardware and software to verification, process, and project management.

Jon won the Volvo-3P Technical Award in 2005, going on to win the 2006 Volvo Technology Award. Jon has secured seven US patents and several international patents. These patents range from multiplexing systems and human–machine interfaces to telemetry systems and driver's aids. Jon has been on the Western Carolina University Master of Project Management Advisory Board and the Forsyth Technical Community College Advisory Board. He has also been a guest lecturer at Wake Forest University's Charlotte, NC campus, and Eindhoven Technical University (Holland).

Jon has authored or contributed to more than 15 books on a range of product development and project management topics. The books he writes are used in bachelor- and master-level classes at universities across the globe, including Eindhoven Technical University, Manchester Metropolitan University, San Beda College Manila in the Philippines, and Tecnológico de Monterrey.

In addition to more than 70 magazines, e-zines, and other outlets, he writes three recurring columns:

1. PMTips Quigley and Lauck's Expert Column
2. Assembly Magazine, Ps and Qs on project management and quality
3. Automotive Industries, Quigley's Corner on automotive product development
4. Thomas Cagley's Software Process and Measurement CAST (SPaMCAST) podcast

Jon has given numerous presentations at technical conferences on various domains of product development, including product testing, learning, agile, and project management. He has also frequently been interviewed by numerous business and project magazines, podcasts, and webinars.

Jon is the co-author or contributed to the books:

1. Co-author of Taylor & Francis book—*Project Management of Complex and Embedded Systems: Ensuring Product Integrity and Program Quality*, ISBN: 1420070256. October 21, 2008.
2. Co-author of Taylor & Francis book—*Scrum Project Management*, ISBN: 1439825157. August 16, 2010.
3. Co-author of Taylor & Francis book—*Testing of Complex and Embedded Systems*, ISBN: 1439821402. December 7, 2010.
4. Co-author of Software Test Professionals/Redwood Collaborative Media Professional Development Series—Saving Software with Six Sigma, ISBN: 978-0-9831220-0-5.
5. Co-author of Software Test Professionals/Redwood Collaborative Media Professional Development Series—Aggressive Testing for Real Pros, ISBN: 978-0-9831220-1-2.
6. Co-author of Taylor & Francis book—*Total Quality Management for Project Managers*, ISBN: 978-1-4398-85055-5. August 28, 2012.
7. Co-author of Taylor & Francis book—*Reducing Process Costs with Lean, Six Sigma, and Value Engineering Techniques*, ISBN: 978-1-4398-8725-7. December 12, 2012.
8. Co-author of Society of Automotive Engineering book—*Project Management for Automotive Engineers: A Field Guide*, eISBN PDF: 978-0-7680-8315-6; eISBN prc: 978-0-7680-8316-3; eISBN epub: 978-0-7680-8317-0. August 31, 2016.
9. Co-author of Taylor & Francis book—*Configuration Management: Theory, Practice and Application*, ISBN: 978-1482229356. April 16, 2015.

10. Contributor to the e-book—*Opening the Door: 10 Predictions on the Future of Project Management in the Professional Services Industry* with Mavenlink and ProjectManagers.net.
11. Co-author, Taylor & Francis book—*Configuration Management, Second Edition: Theory and Application for Engineers and Managers.* July 22, 2019.
12. Co-author, Taylor & Francis book—*Continuous and Embedded Learning for Organizations.* July 2, 2020.
13. Co-author, Taylor & Francis book under contract—*Principles, Processes and Practices of Risk Management.* December 2022.
14. Co-contributor of *Scrum Project Management* topic for the *Encyclopedia of Software Engineering*, ISBN: 1-4200-5977-7; e-ISBN: 1-4200-5978-5.
15. Contributor to the e-book—*The Project Manager Who Smiled* by Peter Taylor. Catalog record for the book is available from the British Library ISBN: 978-0-9576689-0-4, production. June 2013.
16. Co-author of Society of Automotive Engineering book—*Modernizing Product Development Processes: Guide for Engineers*, production. August 31, 2023.
17. Co-author of the Society of Automotive Engineering book—*Dictionary of Testing, Verification, and Validation.* October 30, 2023.

He has co-authored or interviewed hundreds of magazine articles on various product development, teams, and manufacturing topics appearing in over 60 magazines, e-zines, and other outlets (see below).

Jon enjoys the beauty of nature, hiking in the woods, and playing the bass. Jon lives in Lexington, North Carolina. Email Jon at jon.quigley@valuetransform.com.

Wes Chominsky is currently a diagnostic root cause analyzer for a major Class 7 and 8 truck manufacturer in North America. He has 30+ years of experience in the heavy-duty truck industry and has held positions that have ranged from systems integration engineer and systems test and verification engineer in product development to technical support for the aftermarket and manufacturing.

Courtesy of Wes Chominsky.

www.ingramcontent.com/pod-product-compliance
Lightning Source LLC
LaVergne TN
LVHW070827220926
848486LV00027B/29
* 9 7 8 1 4 6 8 6 0 7 8 8 8 *